Landscaping with Edible Plants in Texas

NUMBER FORTY-EIGHT
Louise Lindsey Merrick
Natural Environment Series

Landscaping with EDIBLE PLANTS in Texas

design and cultivation

CHERYL BEESLEY

TEXAS A&M UNIVERSITY PRESS • College Station, Texas

This paper meets the requirements
of ANSI/NISO Z39.48–1992
(Permanence of Paper).
Binding materials have been
chosen for durability.
Manufactured in China
by Everbest Printing Co.
through FCI Print Group

LIBRARY OF CONGRESS CATALOGING-IN-PUBLICATION DATA

Beesley, Cheryl, 1960– author.
   Landscaping with edible plants in Texas : design and
cultivation / Cheryl Beesley.—First edition.
      pages cm—(Louise Lindsey Merrick natural
environment series ; number 48)
   Includes bibliographical references and index.
   ISBN 978-1-62349-321-9 (flexbound : alk. paper)—
   ISBN 978-1-62349-323-3 (e-book)
   1. Edible landscaping—Texas. 2. Plants, Edible—Texas.
I. Title. II. Series: Louise Lindsey Merrick natural
environment series ; no. 48.
   SB475.9.E35B44 2015
   712.609764—dc23
   2015014229

To my grandparents,
who taught me a love of gardening
& the names of plants.

Contents

Foreword, by Jim Kamas   IX

Preface   XI

Acknowledgments   XIII

Introduction   1

Part I. Planning and Preparation   3

Site Evaluation   5

Garden Layout   12

Fencing and Structural Elements   16

Soil Preparation   19

Fertilization   27

Irrigation   29

Cover Crops and Mulches   31

Part II. Design   35

Designing with Edible Plants   37

Traditional Gardens: Parterre and Potager   43

Children's Garden   50

Residential Garden   56

Commercial Garden   59

Japanese Garden   62

Pizza Garden   64

Walled Courtyard Garden   66

Low-Input Edible Landscaping   71

Part III: Edible Plants for Texas   75

Trees   77

Shrubs   125

Perennials   133

Herbs   150

Vegetables   161

Saving Seeds   236

Appendix A: Hybrid Varieties   239

Appendix B: Pest and Disease Control   242

Appendix C: Seed and Plant Sources   266

Index   269

# Foreword

In this long-overdue book, Cheryl Beesley has created a comprehensive guide to the use of edible plants to create landscapes that will prove valuable for both the novice and seasoned landscape designer. Exploring the wide range of landscape uses, this reference logically lays out the steps and considerations that need to be contemplated to create an edible landscape that matches the location and desires of the owner.

Cheryl, a thoroughly qualified landscape designer, combines her in-depth knowledge of the characteristics and care of vegetables, fruits, and herbs with her extensive experience in sustainable horticultural practices to guide in the creation of beautiful living spaces through use of environmentally friendly principles and practices. She clearly lays out the logistical and existing elements of a site that need to be evaluated to create an appealing and productive landscape across the wide range of Texas growing conditions. Her time line for landscape development helps with project planning as well as time and resource budgeting for landscapes that range from simple to elegant. From climatic parameters to soil and water limitations, topography, existing structures, and architectural styles, this guide provides a comprehensive checklist of the numerous factors that will ultimately influence the utility and ascetic appeal of the completed landscape.

Landscaping with Edible Plants in Texas: Design and Cultivation provides clear guidance from evaluating the practical suitability of annual and perennial edible plants—heat, drought tolerance, sunlight preferences—to evaluating the more subtle qualities of seasonal color, form, and texture. It offers a thorough description of plant options, with a comprehensive description of varieties, complemented by an abundance of well-chosen photographs and drawings. Well-thought-out details for annual plant maintenance provide a road map for how to enhance the landscape over time.

I have known Cheryl and her work for a number of years, and it is an honor to write this foreword and have the opportunity to wholeheartedly endorse and recommend this book. Like Cheryl herself, her writing style is completely approachable and enriching. For all who seek to surround themselves with functional and beautiful landscapes, this book will serve as an important resource and testimony to both the art and science of horticulture.

—Jim Kamas
Assistant Professor & Extension Fruit Specialist
Texas A&M AgriLife Extension Service

 # Preface

This book introduces edible plants as a viable option for landscape designers, to bring these plants out of the vegetable garden and into the ornamental landscape. As a gardener and designer, I have always been inspired by the beautiful array of color, texture, and bounty that edible plants have to offer and have never hesitated to mix them in with my ornamental plantings. While walking through the garden one day, I realized that people have land around their homes and places of business they could be putting to use growing edible plants, yet there is a hesitation to mix the two, and the thought of planting a completely edible landscape in an ornamental setting is so removed from the common experience that it is difficult for many to perceive. These plants are lovely and fun to grow, and there is a bonus—they produce edible fruits and vegetables. Although edible plants have commonly been alienated from the ornamental landscape, often hidden from view, they can be a beautiful and unique addition to more visible landscapes, their applications limited only by the imagination of the designer.

Both the commitment and the reasons for landscaping with edible plants will vary from gardener to gardener. A few tomato plants and herb pots will satisfy one gardener, and another will dive into removing the entire lawn and replacing it with edibles during the first season. Beyond the desire to include these beautiful and bountiful plants in the garden, they most likely have shared reasons to include edible plantings, such as spending time outdoors and obtaining food in a more sustainable and less fuel-dependent manner. There is also the certainty of knowing what is going into your food and ultimately your body, while preserving natural resources in a sustainable method.

Perhaps the most important reason to grow an edible landscape is for the love of gardening and to be a part of the miracle of growing a plant from a simple seed to an abundant harvest. *Landscaping with Edible Plants in Texas* is an aid to design and cultivation, intended to assist in keeping plants healthy and productive. Edible plants require more involved care than more traditional ornamental landscapes. The information provided in this book will empower even the novice gardener to have a successful gardening experience and foster a lifelong love of edible landscaping.

✳ Acknowledgments

In many ways, this book has been a labor of love: love of gardening and a desire to share that love with potential gardeners and designers wishing to extend their plant palette and knowledge of landscaping with edible plants in Texas. While the effort that went into this compilation may have been a labor of love, it was a labor nonetheless, and I sincerely thank all the people who helped bring this book to fruition.

First and foremost, I thank my husband, James, without whose consistent support and encouragement this book would have been incomplete. I also thank my parents, siblings, and children, an intergenerational support group and cheerleading team whose belief in this work helped keep me focused.

Second, I thank the following professionals who took the time to review and offer encouragement along the way. I offer my sincere and heartfelt thanks to two women I consider both mentors and good friends: Jill Nokes, landscape restorationist and author, for reviewing the entire manuscript and offering very insightful and constructive critique; and Lauren Renz, registered landscape architect (RLA), for reviewing the designs and helping me "see" what I was missing. Thanks are also in order for Jim Kamas, assistant professor and extension fruit specialist with Texas A&M AgriLife Extension Service, for writing the book foreword, reviewing the section on trees, and offering very detailed and important critique on the content. A special thank you is due to Sam Feagley, professor and state soil environmental specialist with Texas A&M University, for reviewing the information on soils and soil amendments. Thank you also to Jeff Ferris, organic treatment specialist at the Natural Gardener, Austin, for reviewing the information on organic insect and disease control; and to Bob Marie, licensed irrigation specialist, for reviewing the section on irrigation.

Finally, I thank the following gardeners, growers, farmers, orchardists, and homeowners who gave of their personal time and gardening experience to provide real-world experience and application of the design theories presented in this book: Bob Anderson and Gourmet Garlic Gardens; Chef Andre of the Fairmont, Dallas; John and Jimma Byrd of Byrd Pecan Orchard; Jack Dougherty of Bella Vista Olive Orchard; Jo and John Dwyer of Angel Valley Organic Farm; Jim Henry, founding director of the Texas Olive Council; Michelle Mattalino and the Olive Tree Learning Center; Bill McCranie of Chicamaw Farm; Eleanor McKinney, RLA; Leslie and Catherine of Ptarmigan Farms; Trisha Shirey and Stéphane Beaucamp of Lake Austin Spa Resort; Michael and Darlene Starr; and Larry Don Womack and Womack Nursery. I took all the photographs unless otherwise noted.

✸ Introduction

Landscaping with Edible Plants in Texas is a design guide for landscaping with edible plants for use by both professional designers and home gardeners who want to use edible plants in their landscapes, including information about growth habits, form, and other desirable features these plants have to offer. Part 1 provides information about site planning and development. Maps address the regional specifics of Texas' horticultural zones, hardiness zones, regional chilling hours, and average rainfall amounts to help determine varieties that will be most productive for a particular site. Also included is information to aid designers in the preliminary steps of site layout, preparation and construction of organic beds, and options for building fencing with specific information about how to construct the garden elements. Part 2 provides examples of specific design styles and examples of how to use edible plants in Texas landscapes. While each landscape will require unique design responses, these design examples offer ideas and possibilities. As part of the conceptual design, specific plants are selected for their size, form, and preferred growing conditions. Part 3 describes the plant varieties and their specific requirements, guiding designers located in the various areas of Texas. The tree, shrub, perennial, herb, and vegetable sections include cultural information about specific plant families as well as the individual plants and their design uses. Appendices include information on disease and insect control and plant and seed sources. Throughout this book the terms "variety" and "cultivar" are used interchangeably.

Every quality landscape should begin with a good conceptual design, building on the existing elements of the site, analyzing and accentuating the desirable and screening the undesirable features to optimize the site. For professionals who are not designers, it may be tempting to skip the design process and try to improvise a mental vision of the final project. I would not recommend this approach. Even a simple overlay of tracing paper on top of a scaled site plan will reveal patterns and optimal usage of the space. The design section of this book can be a very useful aid in getting the concept on paper for non-designers who may not be accustomed to this process, as well as designers who are not familiar with landscaping with edible plants. The process is not that difficult if specific steps are followed, and the site is much easier to visualize as a cohesive whole on paper. Information on design process, types of design styles, and techniques will impact the success of the overall project.

In addition to design information about edible plants, *Landscaping with Edible Plants in Texas* offers how-to information regarding specific care and variety selections for edible plantings in Texas gardens based on organic principles, including specific information about soil preparation, planting, pruning, insect and disease control, and maintenance to aid the success of the design.

PART I

Planning & Preparation

Front yard with amaranth, Austin, Texas

☀ Site Evaluation

The first step in design is to evaluate the existing site: structures, hardscapes, and plants. Generally, if the property being designed is a residential lot, there is an existing or planned structure or structures that will need to be taken into consideration. Also, the location of the driveway, walkways, and decks is generally determined before the landscape design begins. Although some plants may be transplanted or removed entirely, the designer may have to work around existing plants, especially large trees. Effort spent on the front end of the project in locating these existing elements and committing them to paper in a scaled drawing will save major headaches in the construction phase.

Analysis of the following other existing elements will help in the planting design phase of the project. Not all of these criteria will be relevant to every location, but the checklist will help in evaluating the strengths and weaknesses of the proposed site. See the example of a site analysis that follows.

Sun intensity and pattern: Practically every fruit and vegetable will require plenty of sunlight. Leafy vegetables are able to take a little more shade. Even in Texas, where the sun can be merciless, at least 6 hours of full sun per day will be required for edible landscaping. Of course, the degree and angle of the sun will vary according to the time of year. It is best to become familiar with the sun patterns of the site at all times of the year before deciding where to plant edibles in the landscape.

Wind intensity and pattern: In the northern parts of Texas, the winds are typically from the northwest during the winter. In the more southern counties, wind patterns are affected by coastal weather as well. Generally, winter is the time of year when it is most desirable to block harsh wind. Analyze the winter wind patterns for the site, and plan for an evergreen screen to protect sensitive winter plantings. In the summer, Texans will generally be happy to have any breeze that comes along. With a good design, summer breezes can be funneled into garden and seating areas.

Soil composition: The best investment a gardener can make at the beginning of the design process is a good soil analysis that measures the soil's pH, organic and mineral content, and amount of clay and sand. This is very valuable in determining what amendments are necessary during bed construction to correct imbalances.

Water: Water pH is generally similar to soil pH in a particular area. If your soil has a high pH, it is fairly certain that the water will too. The soil pH can be altered, but little can be done about water pH except to select plants that are suited to the pH of the site. Water availability is a very important consideration for edible gardening. Harvested rainwater is an excellent choice for all plantings. Edible plants require more water than most ornamental plants for good fruit production. An ample, clean water source is imperative.

Drainage: Water takes the path of least resistance, always seeking a lower spot until it is stable. It is very important to study the drainage patterns on a site before designing the location of the beds or other elements that might interfere with surface water flow. It is much easier to work with the existing grade as much as possible and encourage the water to drain into existing swales. These low areas may need to be deepened or widened to accommodate more water. Survey the property after a heavy rain to determine if there is any standing water. It is very important that all planting beds drain properly for the plants' roots to get enough air to thrive.

Views: Plants are a great means for screening or enhancing views. Take into consideration the views from inside as well. Plants and garden structures can be used to draw the eye to a terminus of a view or a sight line. Think also about places on the site and on neighboring properties that might need to be screened from view. Some sight lines will need to be kept open, such as views of children's play areas or pools. It may be desirable to create intimate spaces with plants or structures. A good method for site-view analysis is to sit and stand at various common areas inside and outside the house and consider the desirability of the view and how it might be framed or screened.

Walkways: Accessibility and ease of movement through the site need to be taken into consideration. People will generally take the most direct path to their destination. This does not imply that paths should be straight. Gently meandering paths through garden beds can be aesthetically pleasing. For the analysis, consider different destinations on the site and the routes taken to reach them. The pathway layout will usually become apparent during the design layout. Selection of pathway construction materials will be determined by budget, garden style, and the age and ability of the users.

Utility line locations: It is easy to disregard location of utility lines during the design process. Utility lines are generally underground or above eye level. It is very important to consider where these lines are located to avoid future problems, such as trees growing into overhead power lines or roots choking out septic lines. If there is uncertainty about where the lines are located, have the lines flagged by calling DIG-Tess (811, 800-DIG-Tess, or online at texas811. org).

Intended uses: It is important to analyze a site's usage. Are large grassy areas required for pets or children to play? Is the site used for entertaining large groups or for cozy gatherings? This element of the analysis will determine the scale of the space to be designed, whether it will be divided into intimate rooms or be a large open space.

Storage: Storage and utility requirements are another important consideration. Some gardeners are able to make do with a modest array of tools. Hand pruners and a good shovel are enough for some. Others will have riding mowers, tillers, backpack sprayers, and more, all requiring a place for storage. Consider required tools and a place that will accommodate them comfortably.

Fencing and garden structures: The structural elements of the garden become somewhat permanent, so it is important to carefully think about where to locate them. Views and lot restrictions will factor into this analysis as well. Tall fencing (at least 8 feet) will be required around all edible plantings if deer are in the area, because they will be attracted to an edible landscape. Preventing rabbits, possums, and raccoons will require a tighter mesh fence. Additional information is available in the section on fencing.

Community and home owners associations (HOA) regulations and neighbor compatibility: Many sites are bound to community or HOA regulations. It is very important to become familiar with these rules before designing the garden layout. Fence height and location requirements, plant selection, water restrictions, and setback regulations are common components of the bylaws of muncipalities and associations. Consider the neighbors as well. Although a well-worked compost heap does not have a bad smell, this could become a real point of contention if it is located on your neighbor's property line.

Regional Conditions

An additional consideration in site analysis for Texas gardens is the specific regional conditions where the site is located. The maps provided here will help in the site analysis process.

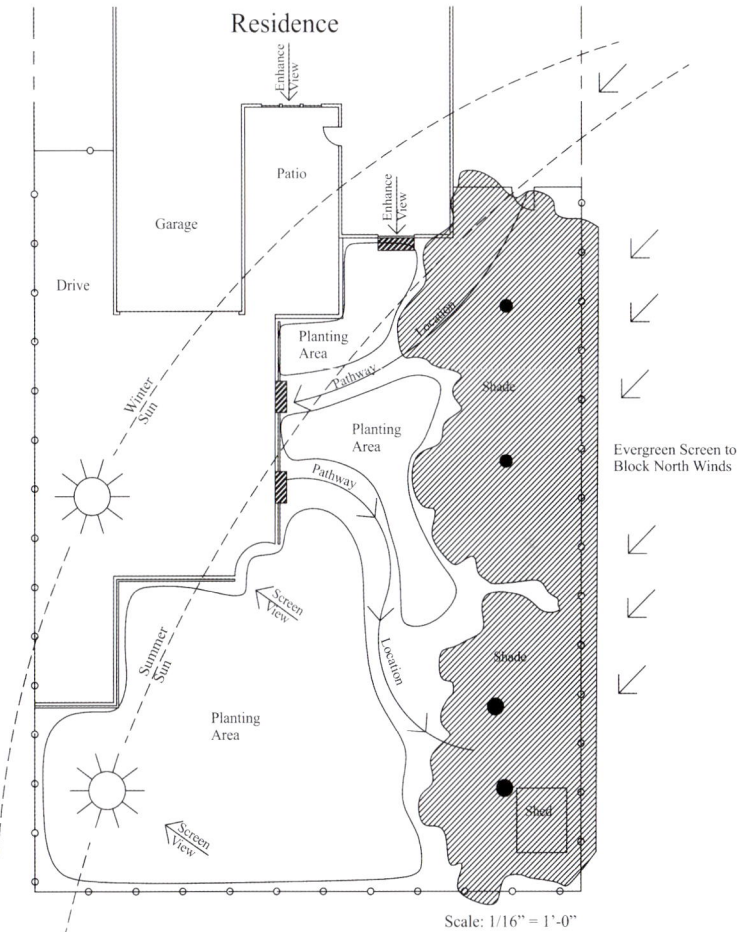

Site analysis

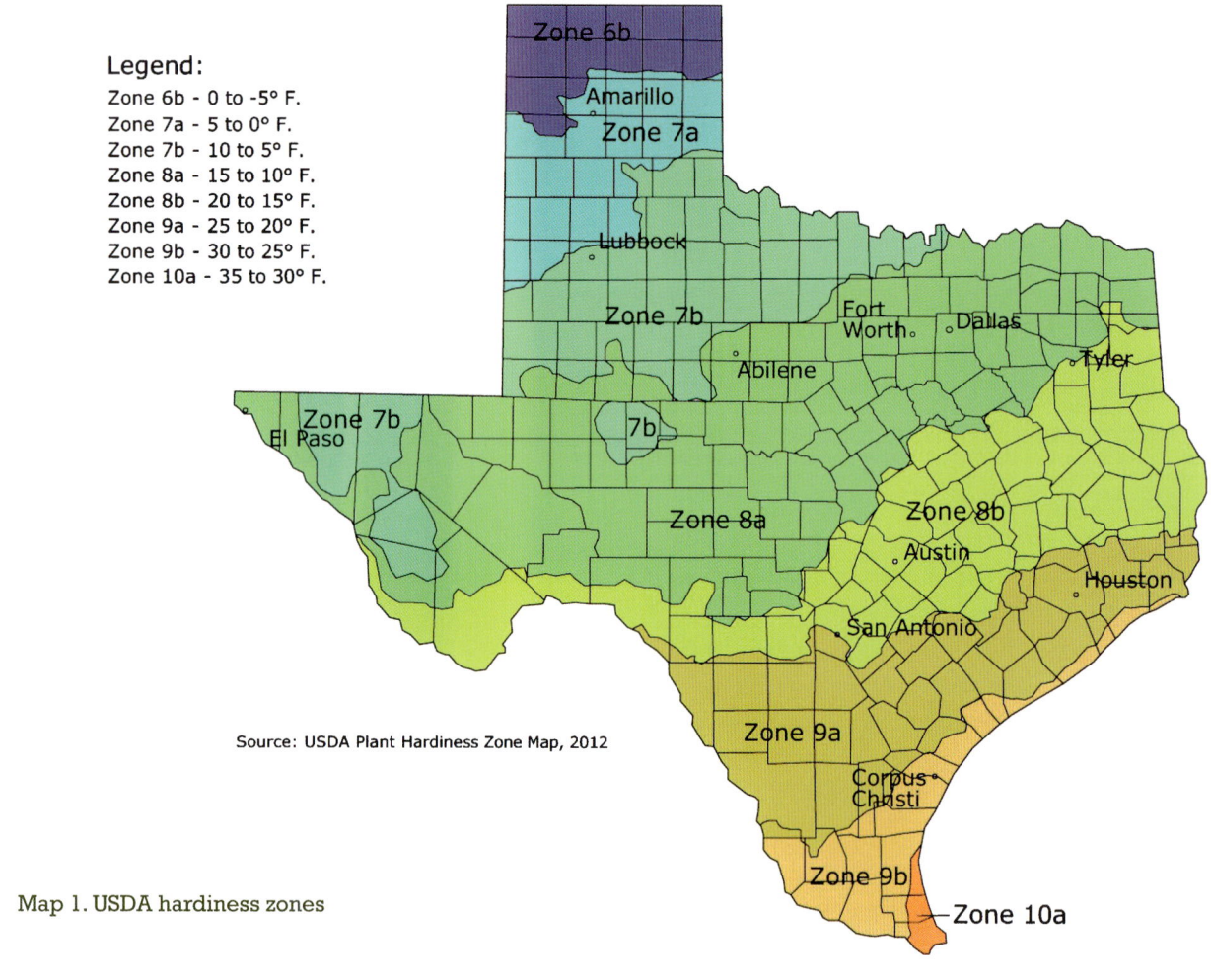

Legend:
Zone 6b - 0 to -5° F.
Zone 7a - 5 to 0° F.
Zone 7b - 10 to 5° F.
Zone 8a - 15 to 10° F.
Zone 8b - 20 to 15° F.
Zone 9a - 25 to 20° F.
Zone 9b - 30 to 25° F.
Zone 10a - 35 to 30° F.

Source: USDA Plant Hardiness Zone Map, 2012

Map 1. USDA hardiness zones

USDA HARDINESS ZONES

It is important to know in which hardiness zone the garden is located to determine if a particular plant is suited for a site. For example, loquats may not produce reliably north of zone 8b. Most of the plants in this book are well suited for all Texas gardens. Information on specific hardiness requirements can be found in the plant descriptions.

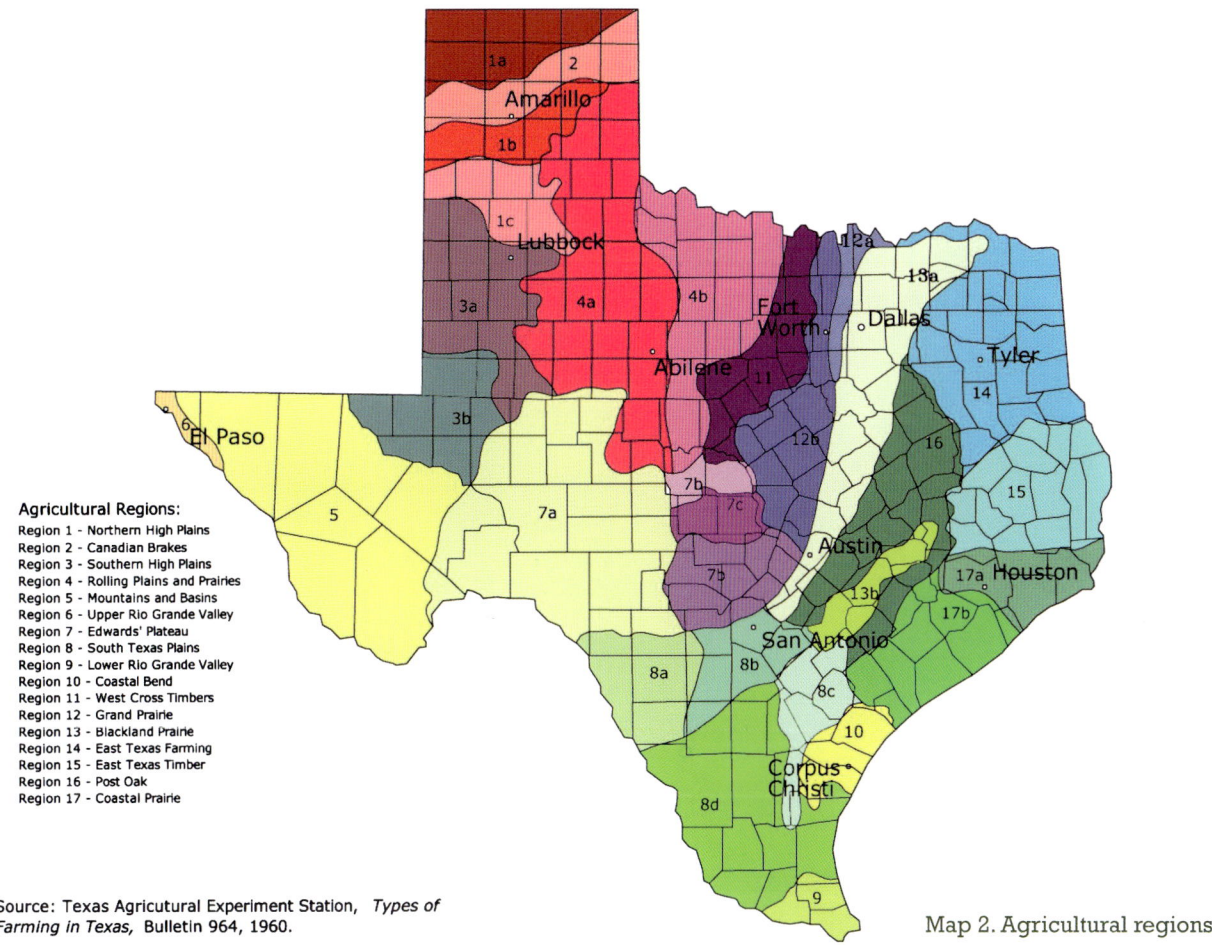

Agricultural Regions:
Region 1 – Northern High Plains
Region 2 – Canadian Brakes
Region 3 – Southern High Plains
Region 4 – Rolling Plains and Prairies
Region 5 – Mountains and Basins
Region 6 – Upper Rio Grande Valley
Region 7 – Edwards' Plateau
Region 8 – South Texas Plains
Region 9 – Lower Rio Grande Valley
Region 10 – Coastal Bend
Region 11 – West Cross Timbers
Region 12 – Grand Prairie
Region 13 – Blackland Prairie
Region 14 – East Texas Farming
Region 15 – East Texas Timber
Region 16 – Post Oak
Region 17 – Coastal Prairie

Source: Texas Agricultural Experiment Station, *Types of Farming in Texas,* Bulletin 964, 1960.

Map 2. Agricultural regions

AGRICULTURAL REGIONS

Although this map is based on agricultural growing regions, these regional designations are associated with soil types, which impact the specific soil amendments needed to improve the soil conditions. For example, soils in the Blackland Prairie are generally heavy clay, and soils in the Edwards Plateau are often a very thin layer of high-pH topsoil over limestone. While compost can help improve both of these soil types, it is important to know the region where the site is located to determine the best soil amendments. For example, this information would guide the reader to know whether to add haydite to help loosen clay soils or to add K-Mag to soils with a limestone base to increase acidity.

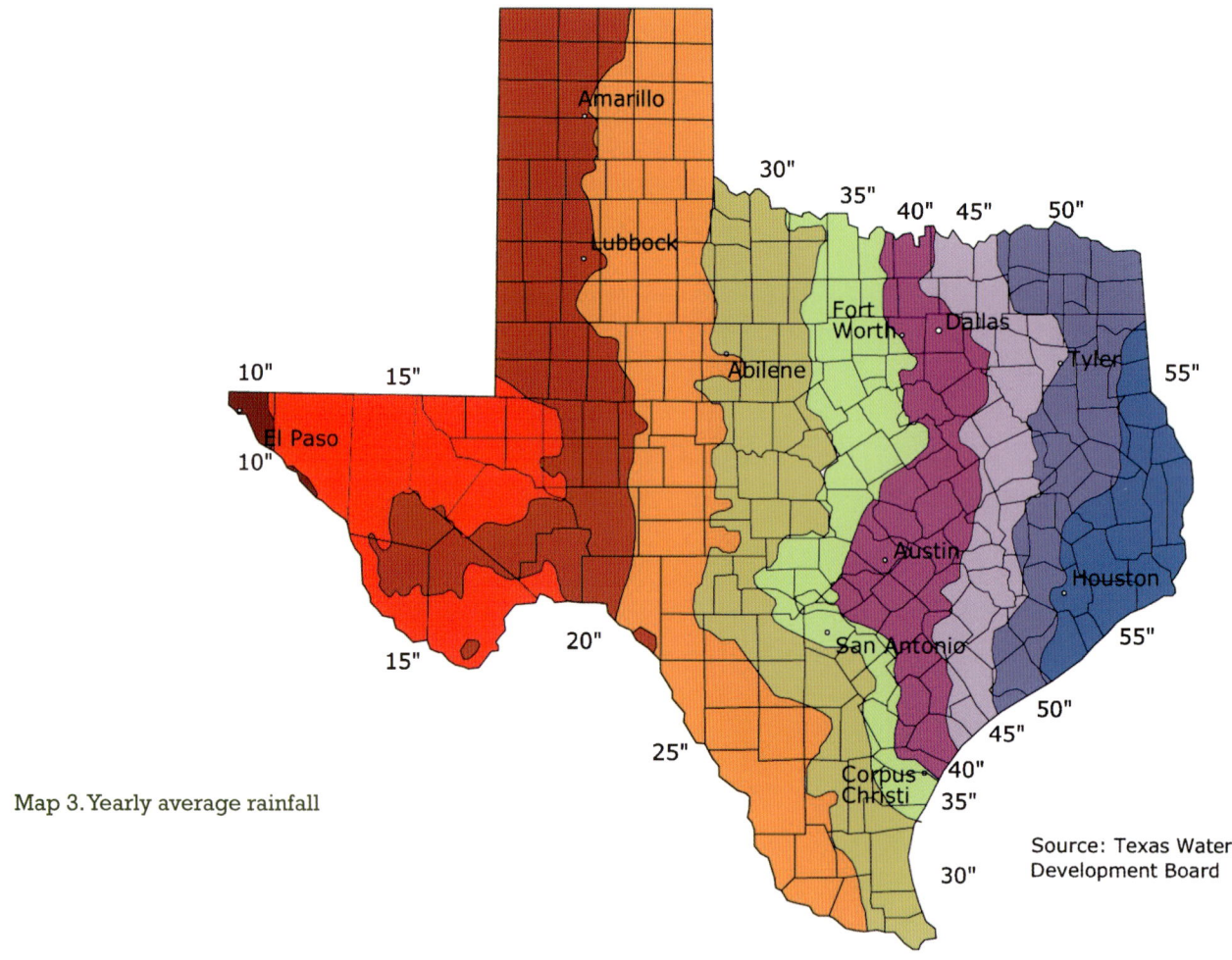

Map 3. Yearly average rainfall

Source: Texas Water
Development Board

AVERAGE ANNUAL RAINFALL

Average annual rainfall amounts will decrease the farther west a garden is located in Texas. For example, a garden in El Paso receives an average of only 10 inches of rain annually, while a garden in Houston receives an annual total of 55 inches.

When planning for a garden's water requirements and selecting which plants would do best with the available rainfall, it is helpful to know how much rain is available on average and when to plan for water collection and conservation if necessary.

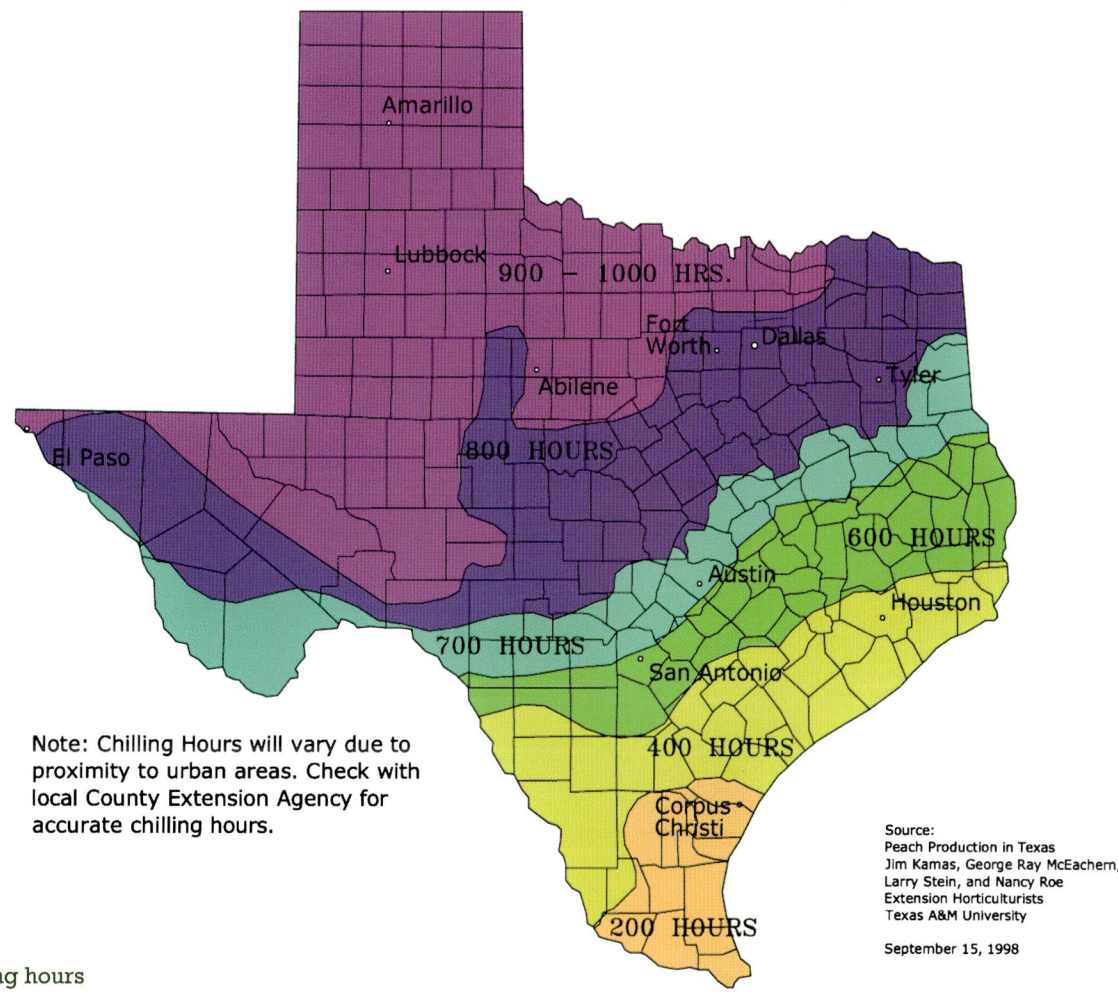

Note: Chilling Hours will vary due to proximity to urban areas. Check with local County Extension Agency for accurate chilling hours.

Source:
Peach Production in Texas
Jim Kamas, George Ray McEachern,
Larry Stein, and Nancy Roe
Extension Horticulturists
Texas A&M University

September 15, 1998

Map 4. Chilling hours

CHILLING HOURS

Chilling hours are the number of hours of winter chilling required to break down growth inhibitors for flowers and buds, generally at temperatures between 45°F and 32°F. Most fruit trees have specific chilling-hour requirements dependent on variety. The chilling hours map can help determine which varieties will produce most effectively for the site location.

☀ Garden Layout

Once a careful analysis of the site has been done and the existing elements are drawn to scale on a base plan, the design process continues with the addition of the new elements. When considering a vegetable garden, it is easy to think of the traditional row arrangement, which is based on a farming model that lends itself to more convenient crop cultivation and harvesting. In the urban landscape, it is not necessary, or even desirable, to replicate this linear design. Typical residential and commercial properties lend themselves to a less rigid, more interesting layout. Generally, garden

Pear tree allee east of Hochosterwitz Castle in Austria (Photo by Johann Jaritz)

layout is limited only by the elements discussed in the previous section on site analysis. Beyond these factors, the site is open to the designer's imagination.

Design Considerations

Taking the following elements into consideration will aid the designer in determining bed layout and plant selection.

Style of existing architecture: What is the architectural style of the structures near the garden? If modern, the bed design may be more angular and plants may be laid out in a grid pattern, with interesting repetition of plant patterns. If the architecture is French colonial, a kitchen garden with formal parterres may be the appropriate design style. Beautiful enclosed courtyards can be planted with edibles around a Spanish colonial house.

Scale: The size of the site will dictate the quantity and ultimate size of the selected plant material. A large pecan tree may be out of scale with a zero-lot line condominium, and dwarf fruit trees may be lost in a large, rambling property. It is important to keep in mind the desired views and how the plants will appear from a distance. Comfortably sized outdoor "rooms" should not feel suffocating, however; there should be a comfortable feeling of enclosure. It may be helpful to place the plants in containers on top of the ground before planting and sit in the space to determine if the space should be smaller or larger.

Ultimate size of plants: It is very easy to be overzealous and place too many plants in an area. Keep in mind the ultimate size of the plants. Draw the trees, shrubs, and perennials to scale on the plan to be certain to plant the accurate number at the correct spacing.

Desired structural elements: Trellises, arbors, fountains, and walls should be drawn on the planting plan, as these items will be placed in semi-permanent locations and will determine the location of planting beds and walkways.

Types of fruits and vegetables: The easiest method for deciding which edible plants to include in the garden is to make a list of the fruits and vegetables enjoyed by the people who will be eating from the garden. It is also important to plant these fruits and vegetables in appropriate quantities. If okra is eaten for only two or three meals a year, it is probably not a good idea to plant fifteen plants.

Ease of access: Edible plants require quite a bit of care, and it is important to be able to reach them. Bed widths should not be more than a person's arm can comfortably reach. Generally, beds that are open on only one side should not be more than 3 feet in width, and beds that are open on two sides should not be any wider than 6 feet.

Plant design: Plants in the edible garden create interest with color, texture, and form. The first consideration should be which plants are going to be placed in permanent locations. Site the trees, shrubs, and perennials on the plan, making sure they have enough space to reach their mature sizes. The spaces left in the beds after the permanent plantings can be filled with annual plantings, typically vegetables and flowers.

After the edible garden has been designed on paper, it is a good idea to consider if the plan will be implemented all at once or over a period of time. A large-scale edible landscape is very time consuming. What looks manageable on paper can be quite overwhelming in reality. The first step in any good gardening project is a careful evaluation of available time. It is much better to have a few fruit trees and some pots of herbs or an allotment in a community garden to begin with and expand to more adventurous gardening projects as time and enthusiasm allow. The following 4-year plan is an example of how a garden design might be implemented.

After careful evaluation and planning, the design solutions should become apparent in the design process. The site design shown here offers one solution for the previous site analysis plan.

Additional Ideas for Urban Gardens

If space is limited, other options are available for landscaping with edible plants.

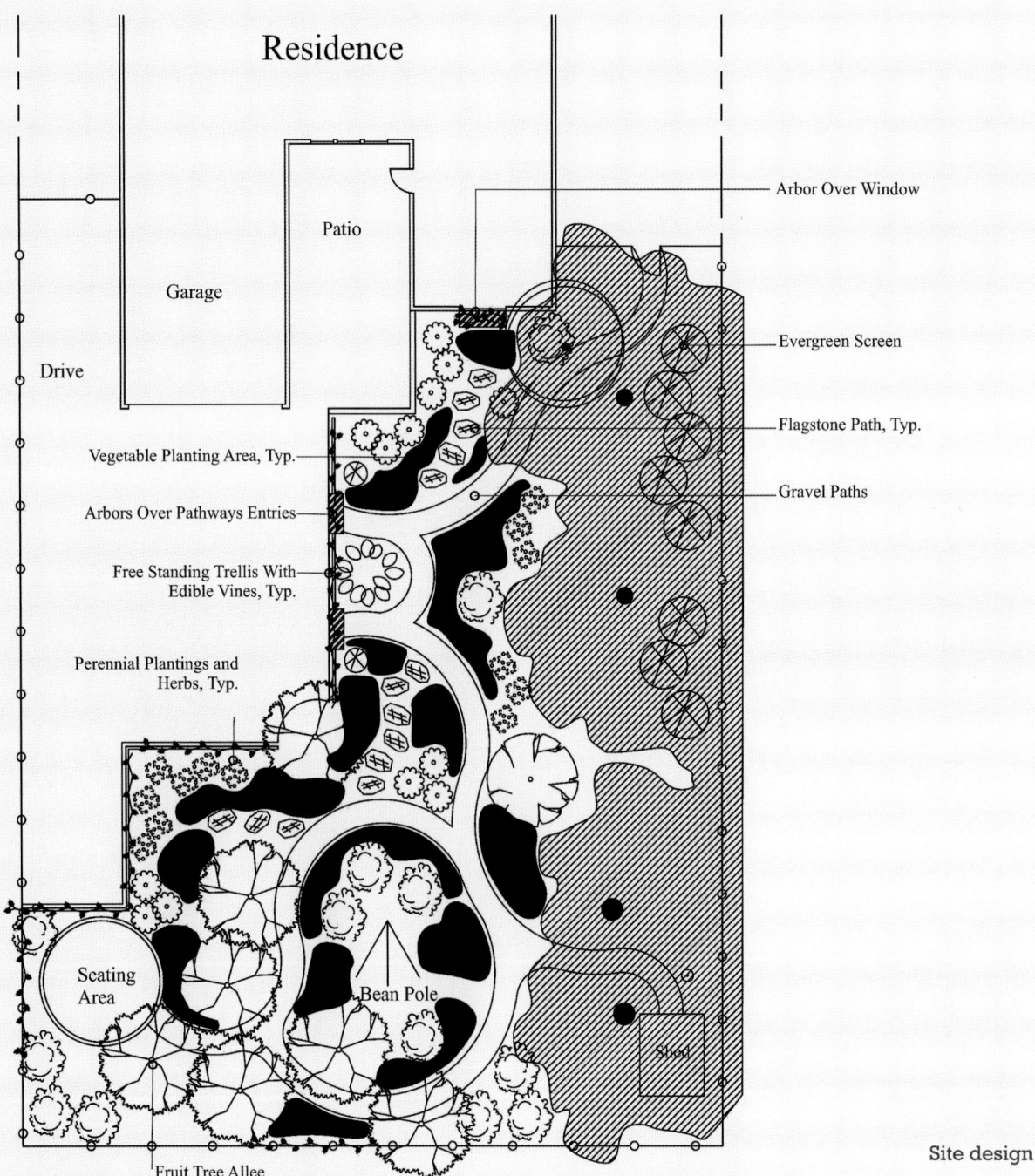

Residence

Garage

Patio

Drive

Arbor Over Window

Evergreen Screen

Flagstone Path, Typ.

Gravel Paths

Vegetable Planting Area, Typ.

Arbors Over Pathways Entries

Free Standing Trellis With
Edible Vines, Typ.

Perennial Plantings and
Herbs, Typ.

Seating
Area

Bean Pole

Shed

Site design

Fruit Tree Allee

Scale: 1/16" = 1'-0"

Year 1: Prepare the evaluation and design of overall landscape. Test soil to determine required amendments. Prepare areas for trees, which take the longest time to mature and produce fruit. Clear all existing vegetation from areas where trees will be planted, research variety selections, and be prepared to order the trees in plenty of time to have them delivered at the best time for planting. Adjust irrigation systems to include bubbler emitters for each of the trees. Begin to prune fruit trees. (See pruning information in part 3.)

Year 2: Construct hardscape elements, such as raised bed frames, grape and bramble fruit trellises, compost heaps, fences, arbors, and cold frames. Continue to prune fruit trees.

Year 3: Prepare bed areas, adding amendments according to results from the soil test. (See the section on soils for more information on amendments.) Because perennials will be in the same location for many years, it is important to pay special attention to perennial bed preparation. Continue to prune fruit trees. Lay drip line for all beds.

Year 4: Plant perennials and annuals in season.

Neighborhood and community gardens: Check with neighbors about connecting gardens encompassing more than one lot. This creates an interconnected ecology so that a wider variety of fruits and vegetables can be produced. One neighbor might grow corn; one, green beans; and another, tomatoes. There will be plenty to go around for everyone, and an individual does not need to worry about canning 15 bushels of tomatoes. In addition to efficient production, neighborhood gardens are a great way to get to know the neighbors. Community gardens are another way to build lasting friendships while growing healthy produce and enjoying fresh air and sunshine.

Balcony and rooftop gardens: Even apartment dwellers usually have room to grow some of their own produce. Fresh herbs are expensive in the grocery store, and they are very easy to grow in pots on a balcony as long as they receive at least 4 to 6 hours of sun per day. Container-grown dwarf fruit trees produce delicious full-sized fruit in small spaces. Patio and cherry tomatoes, peppers, eggplant, and edible greens are good choices for containers. Consider upright elements to train vining plants to get more growing area in small spaces and create "living walls."

Apartment community gardens: With the permission of management, apartment complexes offer a wonderful opportunity for communal vegetable growing. It is a great way to meet neighbors and build friendships while utilizing common areas to produce healthy fruits and vegetables.

Urban farms: Many urban farms enjoy sharing their produce in exchange for a few hours of help with the weeding and harvesting. Some housing development projects have designed urban farms into their communities, where homeowners are given a share of the produce as a part of homeowners' property dues.

Fencing and Structural Elements

It is important to consider the types and location of fences and structures during the initial design layout. There are primarily two types of fencing that will relate to the edible urban garden: perimeter fencing to keep out foraging animals (mostly deer, rabbits, and raccoons) and decorative fencing to enhance the garden and create opportunity for vertical planting. Unless a solid fence is used, it is not practical to use perimeter fencing for plant supports, because the animals will eat the plants through the fencing.

There are many types of fence and structure materials; wrought iron, bamboo, wood, wire mesh, plastic mesh, and masonry are the most common. The selected material will depend on certain factors, such as style of architecture, types of pests in the area, and budget.

Gourds trained on overhead arbor (Photo by Andrew Edmonson)

Perimeter Fencing

The following are some options for fencing to deter foraging animals.

Deer fencing: It is generally recommended that a fence meant to keep deer out of the garden should be at least 8 feet tall. Deer are able to jump shorter fences. This does not have to be a terribly expensive investment. There are effective plastic mesh products available that are able to keep these ravenous intruders at bay. The size of the mesh is generally between 1¾ and 4 inches square. This fencing is not meant to be decorative and is described as "invisible." Of course, it can be seen, but it is not overly obtrusive. A more decorative fence may be used. Wire mesh attached to weather-resistant wood posts is very attractive. The type of fencing is not as important as the height and stability when dealing with deer.

Rabbit fencing: An effective fencing material to prevent rabbits from getting into the garden is 36-inch chicken wire. Rabbits will not generally jump over fences, but they will burrow under, so it is important to bury the fencing. Dig a trench 6 inches deep and 8 inches wide along the base of the fence location. Then bend the chicken wire into an "L" shape with the base of the "L" lying along the bottom of the trench. Pound stakes into the ground on the garden side of the trench to hold the height of the chicken wire. After the fencing is attached, fill the trench.

Raccoon fencing: Raccoons particularly love corn and melons. Regular fencing may not be sufficient to keep these clever creatures out of the garden. They are able to climb most types of fences with their fingerlike paws. Raccoons can be fooled with a "C"-shaped fence made of chicken wire 48 inches high. The bottom 15 inches of the "C" rest on the ground, and the middle 18 inches are secured to stakes, allowing the top 15 inches to flop over (away from the garden). The little bandits will not be able to climb over the arching chicken wire.

The most effective method for keeping raccoons from raiding the garden is to string electric fencing. Be sure to use a low-voltage electric line with a pulsating current. Continuous current electric lines are hazardous for birds.

Structural Elements

Garden structures are beautiful and useful garden elements to add vertical production space, screen views, and create more intimate outdoor garden spaces or "rooms." Plants trained to grow on a garden structure create a beautiful upright element and will benefit from the added support, sun, and air circulation. Aside from fencing, there are three main types of garden structures: trellises, arbors, and pergolas. These three forms can be constructed in a versatile array of styles and from numerous material types to suit many applications.

Trellises: These structures may be free-standing or attached to a wall. They are generally grids or lattice attached to frames, although they may be fan-shaped or designed in artistic free-form patterns. Free-standing trellises are supported by posts and can be used to separate garden spaces or to serve as dramatic garden accents. Many trellises in the edible garden are simple structures to keep plants off the ground. These structures may be simple tomato cages made from rounds of mesh wire, or they may be a tepee-type structure. Just because they are simple constructions does not mean these arrangements cannot be very dramatic garden elements, especially if used on a grand scale or in a repetitive pattern.

Arbors: Arbors differ from trellises in that they have an overhead element. They may be used in many ways, such as entryways into separate garden spaces, as accents over doorways or windows, or as a way to produce shady seating areas. An interesting method for creating a living arbor is to plant fruit trees at the corners of the arbor. When they are taller than head height, train branches to grow over the roof so they eventually meet in the middle, or the branches can be intertwined to produce a roof. Fruit trees may also be grown in the same way with structural elements for support, allowing the fruit to hang down between the crosspieces. It is important to use standard-sized fruit trees for overhead elements, as dwarf trees will probably not grow tall enough to drape over arbors.

Pergolas: Pergolas are arbors laid out in a colonnade form with symmetrically spaced posts. Pergolas are generally used in the same way as arbors but are better for a larger area.

Growing area for edible plants can be greatly increased by supporting plants on vertical elements. Plants grown on supporting structures are also less susceptible to fungal problems, as the vegetation is not in direct contact with soil, which harbors many fungal pathogens. This type of growing also allows for more air circulation around the plants to prevent water from sitting on the leaves (another cause of fungal disease).

Some plants must be grown on supporting structures to do well; others will do fine growing on the ground, but they may be grown on vertical supports for dramatic effect. The following is a list of suggested plants for trellising; the ones with asterisks will require support for optimal production.

- **Apple and pear** (dwarf varieties for trellising, standard to create living roofs)
- **Blackberries** (Some varieties require trellising; others are free-standing.)
- **Cucurbitaceae family:** Cucumbers*, squash, and melons (Suspend heavier fruit in slings made from cheesecloth, muslin, or old pantyhose.)
- **Fig**
- **Grapes***
- **Leguminosae family:** Beans*, peas* (Pole beans and peas require staking; bush varieties do not.)
- **Loquat**
- **Solanaceae family:** Tomatoes*
- **Stone fruits:** Nectarines, peaches, and plums (dwarf varieties for trellising, standard to create living roofs)

☀ Soil Preparation

The importance of good soil preparation cannot be overstated. It is unarguably the most important step to ensure the health and productivity of the plants. Organic plant care is based on good soil preparation. Healthy soils produce healthy plants that are more able to withstand diseases and insect infestations. The humus-rich soils are more able to allow plant roots to reach deeply into the ground to feed and anchor healthy aboveground leaf growth. It takes a lot of muscle to prepare the soil and optimize soil fertility. The results are worth the time and effort. These practices employ nature's design for nurturing plants with a little help from the following organic products.

There are more than twelve hundred soil series in Texas, and it is not uncommon for several of them to appear in the same yard. Complicating things even further, most yards have soil brought in from surrounding areas as part of the site construction. As anyone who has driven across Texas can tell you, Texas is really big! The major land resource areas, illustrated in map 2, are a very rough approximation of the soil diversity found across this wide area. The deep sandy loams of East Texas that can support the growth of large pine trees is very different from the mainly shallow, calcareous soils of the Edwards Plateau. Most regions have a combination of soil types, and a soil test is valuable in determining the proper amendments to correct soil deficiencies.

There are twelve soil textures that are different combinations of sand, silt, and clay. Soil "tilth" refers to the ease of workability of the soil. Clayey soils have very fine particles with smaller pore size that can constrict the flow of air and water. Sandy soils have much larger pore spaces and allow the water and nutrients to move through more easily. Silt is in between the size of sand and clay particles. It is important for edible plants to have a loose, friable soil with plenty of organic matter for the optimal movement of air and water between the soil particles. A well-structured soil has stable soil aggregates made up of a combination of sand, silt, clay, and organic matter that allows a plant's roots to spread and grow deeply, increasing the roots' surface area and ability to uptake nutrients and water.

The soil pH and mineral content are determined by the bedrock geologic layer, called parent material, from which it originates. Wind, rain, temperature, plants, and animals work over many years to break down the rock into finer particles. Most soils in Texas are derived from a mixture of small particles of sand, silt, and clay that have been transported by wind and rain. These particles are mixed with decomposing organic matter

Prepared beds, Natural Gardener, Austin, Texas

through physical, chemical, and biological processes to form soil. The time it takes to produce an inch of soil from easily weathered rock under natural conditions is estimated at anywhere from 500 to 1,000 years. Generally the soil contains the elements a plant needs to be healthy, but these elements may be unavailable to the plant if the soil is not well structured. The pH of the soil and the solubility of the nutrients are additional factors in nutrient availability. A well-prepared soil will be high in organic matter that is full of microbes. A healthy soil will contain between 1 million and 100 million microorganisms in one gram of soil (about the mass of a regular paper clip). Stated in another way, there are an estimated 45 billion microbes in only a pound of healthy soil. Obviously, it is impossible to see them without a microscope, but the results of their handiwork are obvious, as seen in the rich humus left behind.

Roots have developed symbiotic relationships with certain soil microorganisms. Beans, peas, and other legumes have a particular relationship with specific nitrogen-fixing bacteria called rhizobia, which are housed in nodules in the plants' roots. Thus, legumes are often used as cover crops between desired crops. Beneficial mycorrhizal fungi work in a symbiotic relationship with all types of plants to help their roots obtain difficult-to-extract nutrients from the soil. The fungal strands are very tiny, even smaller than the plants' roots to which they are attached. They are able to reach nutrients that would otherwise be inaccessible.

Compost Heap Construction

Fortunately, there is a wondrous soil amendment that is able to remedy most soil problems and increase beneficial microbial activity. That soil amendment is compost. It can be worked into heavy clay soils to help form aggregates to improve soil structure over time, which will increase the number and size of pores between particles so water and air can move more freely through the soils. Compost worked into sandy soils improves aggre-gate stability and water-holding capacity. Properly prepared compost is rich in the nutrients that plants require for healthy growth and interacts with the nutrients already in the soil to make them more available. Composting is a biological process that converts organic matter into humuslike material suitable for use as a soil amendment and organic fertilizer, combining decomposing vegetative matter with manures to produce a compost rich in nutrients and humus. Examples of this process in nature can be seen on the forest floor as the layer created below fallen leaves where this sweet-smelling, crumbly humus is found. Humus, teeming with microbes, is key to good soil production.

Although compost is formed through biological processes, making compost is not very technical. Decaying plants and manure will break down and release nutrients without any human assistance. Nothing more complex than throwing kitchen vegetable scraps, eggshells, and coffee grounds on top of a pile of leaves and covering them with more leaves or yard trimmings will produce usable compost; however, there is a "recipe" that will produce a higher-quality finished product. Layer leaves, grass clippings, and yard trimmings with vegetable scraps, coffee grounds, and crushed eggshells at a ratio of 3:1, respectively. Sprinkle these layers with a fine layer of soil from the yard to inoculate the heap with local soil microbes. Water the layers with rainwater if it is available; otherwise, water from the hose will work. Do not allow the compost to become soggy. Poor aeration can cause the heap to become anaerobic and prevent it from breaking down properly. It is good to

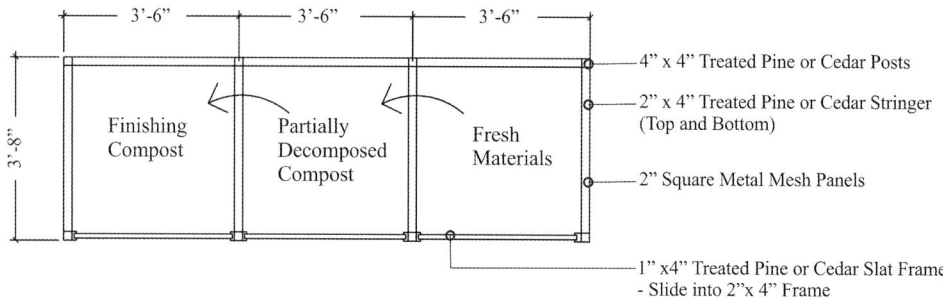

3'-6"   3'-6"   3'-6"

3'-8"

Finishing Compost

Partially Decomposed Compost

Fresh Materials

4" x 4" Treated Pine or Cedar Posts

2" x 4" Treated Pine or Cedar Stringer (Top and Bottom)

2" Square Metal Mesh Panels

1" x4" Treated Pine or Cedar Slat Frame - Slide into 2"x 4" Frame

Plan View

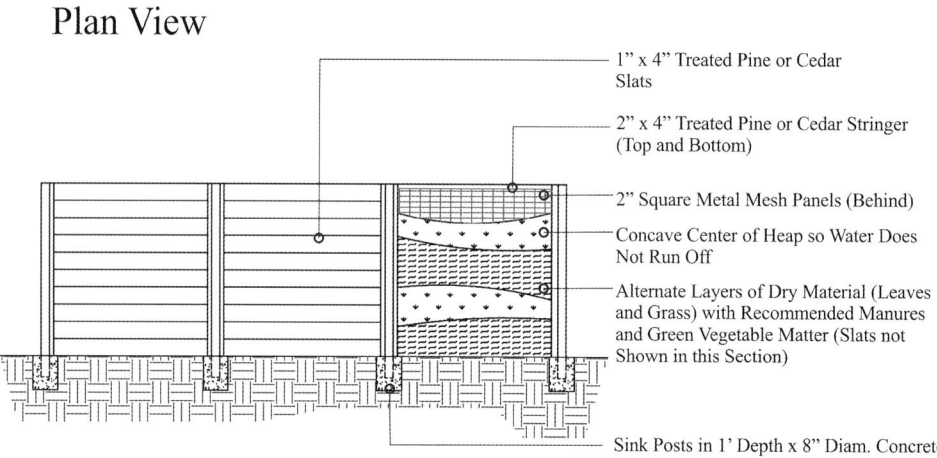

1" x 4" Treated Pine or Cedar Slats

2" x 4" Treated Pine or Cedar Stringer (Top and Bottom)

2" Square Metal Mesh Panels (Behind)

Concave Center of Heap so Water Does Not Run Off

Alternate Layers of Dry Material (Leaves and Grass) with Recommended Manures and Green Vegetable Matter (Slats not Shown in this Section)

Sink Posts in 1' Depth x 8" Diam. Concrete

Elevation

Compost heap construction (NTS)

aerate the heap by turning the ingredients, which causes them to break down more quickly. Turn the pile every 2 to 4 weeks or when its temperature starts to decrease. Temperature can be determined with a compost thermometer. Optimal temperature range is between 135°F and 160°F when the compost heap is most active. Temperatures will decrease as the material breaks down. Although weed seeds should not be put in a compost heap, they can be carried with the wind and on animals.

Hotter compost heaps will help kill the weed seeds. It is also very important not to put diseased plant material in compost heaps.

It is not uncommon for pill bugs and sow bugs, little armadillo-like insects, to be very prolific in compost heaps. Ants can also be a problem. These little critters are not as big a problem in heaps that are higher in temperature. These insects are not harming the pile but are actually helping break it down. It is not good to incorporate them into your garden though. If infested compost is ready to be put in the garden, the material can be spread out on a tarp to heat up, and the pests should depart.

Manures used for compost heaps are mostly cow and horse manures usually available from local dairies and farms, garden centers, or bulk soil suppliers. Do not use cat or dog feces for compost. Chicken and rabbit manure may be used, but they are much higher in nitrogen and phosphorus. Finished humus will have an approximate carbon to nitrogen ratio of 10:1, no matter what process is used to decompose the ingredients. For ongoing compost production, it is a good idea to keep two or three bins, alternately adding materials to the unfinished bin while another is left to "cook."

Additional Soil Amendments and Fertilizers

Soil amendments are used in the initial bed preparation and should be followed by a regular fertilization program. Some of the following products are followed by three numbers in parentheses. This is the guaranteed fertilizer analysis representing the percentage of nitrogen, phosphorus, and potassium (N, P_2O_5, and K_2O) in the product. The numbers given for the following fertilizers are guides and may vary by manufacturer. Many of these products are amendments used for specific purposes but have fertilizer formulations as well. Organic sources of nutrients release slowly as the microorganisms in the soil break down the chemical composition into a usable form.

Agricultural limestone: Low pH can hinder the availability of nutrients for non-acid-loving plants. Calcitic limestone or dolomite limestone (which has magnesium) can help improve soils with low pH (less than 5.5 for non-acid-loving plants; 6.2 for legumes). A soil test is recommended to determine the amount of limestone to apply. Low soil pH is not generally a problem for Texas gardeners, except in certain areas of East Texas. Typically, if there is a need for limestone application, this application will be repeated every 3 to 5 years.

Alfalfa meal (3-2-2): This is a good source of plant essential nutrients. Mix in with existing soil at time of planting and top-dress plants yearly.

Bat guano (8-4-1): Guano is a concentrated soil amendment that contains nitrogen, phosphorus, and potassium as well as trace minerals, which is an excellent amendment at planting time. Add 1 or 2 teaspoons for each plant set or 1 to 2 pounds per 100 square feet.

Blood meal (13-1-0): This product is a good source of nitrogen. A little blood meal goes a long way, so follow container directions carefully.

Bone meal (4-12-0): This is a slow-release soil amendment used to increase phosphorus in soils with a pH of 7 or lower. Phosphorus is an element necessary for flower production. Work in around the existing soil at the roots.

Corn gluten meal (9-0-0): This product is usually thought of as an organic herbicide for broadleaf weeds. It is a very good source of nitrogen as well. Corn gluten meal should not be used in beds where seeds are to be sown or have recently emerged.

Cottonseed meal (6-1-1): This is a by-product of vegetable oil production that is a good slow-release amendment and may acidify soil. Because it has a high nitrogen content, cottonseed meal encourages foliage growth, so it is best to add after flower and fruit production.

Earthworms: These soil builders are not officially "soil amendments," but they are a great addition at the time soil is being prepared. These little dirt movers develop soil tilth and increase nutrient availability.

Earthworm castings (3-1-1): This highly nutritive soil amendment is derived from the "leavings" of the earthworm. It is a good source of plant nutrients, including nitrogen, phosphorus, and potassium as well as calcium, iron, magnesium, sulfur, and a large number of trace minerals. Earthworm castings are an especially good amendment to apply at planting time. A handful mixed into the planting hole will provide a good start for young seedlings, or castings can be dusted over freshly planted seedbeds. Earthworm castings can also be added to container soil mixes.

Expanded shale (haydite): Expanded shale, an inorganic soil amendment that improves soil drainage and air circulation, can absorb about 20 percent of its weight in water. This product can be used as a lightweight soil amendment for planters and rooftop gardens where a lighter product is preferred.

Feather meal (7 to 12-0-0): This soil amendment is a by-product of the poultry industry and is a good source of nitrogen.

Fish meal (9-4-0): This soil amendment is made from dried fish waste. It is a source of nutrients that promote healthy root, flower, and foliage growth.

Granite dust: This soil amendment is a by-product from granite quarries that provides many nutrients, such as potassium, calcium, magnesium iron, manganese, copper, and zinc.

Gypsum: This soil amendment is a good source of calcium and sulfur. It is used to aid in water infiltration, to reduce the effects of soils with high sodium levels and the effects of aluminum toxicity in low-pH soils. Gypsum can cause leaching of essential nutrients and needs to be applied on a regular basis.

Kelp meal (1-0-1): This product is good to apply at the time of planting. It is an excellent root stimulator and microbial stimulator that helps prevent transplant shock, boost growth, and prevent fungal problems.

Langbeinite: This mineral contains potassium, magnesium, and sulfur.

Oyster shell calcium: Soils with low pH can benefit from this soil amendment. It improves nutrient uptake by increasing the soil pH.

Rock phosphate (0-3-0): This extremely slow-release source of phosphorus for soils with pH values less than 5.5 has a particle size similar to that of talcum powder. Mix in around roots at time of planting, and top-dress plants yearly, usually in the fall for spring flowers.

Sulfur: Add sulfur to lower soil alkalinity and make iron, zinc, copper, and manganese more available. Apply at the time of soil preparation at recommended rates to lower soil pH. The fine dust should not be inhaled. Any soil amendment to lower the pH should be applied based on a soil test. Applying too much sulfur could lower the pH too much, perhaps to 2 or 3 within a month. Texas A&M AgriLife Extension Service Soil, Water, and Forage Testing Laboratory does not recommend acidifying agents be added to soils that contain carbonates (calcareous limestone soils) because in these soils it is very difficult to determine which part of the carbonate is reactive with the acid and which particles are too large to be reactive. They recommend that plants be selected that grow well in alkaline soil conditions.

Of course, not every soil will need all of these soil amendments. It is a good idea to get a soil test to determine soil pH and nutrient recommendations before adding any source of nutrients. If a problem is detected, soil tests should be done yearly for 3 to 4 years or until there is very little change in the nutrient concentrations in the soil. Soil testing should be done every 3 years thereafter. The soil test recommendations will be usable if no changes are made to what is being planted or to sources of nutrients. Certain problems such as cotton root rot should be included in the testing in areas where this disease is common, as there is no treatment and some plants are highly susceptible. The following labs offer information on their websites about how to collect and send samples. The Texas A&M University Soil Characterization Laboratory website can offer further information about soil types and amendments: http://soildata .tamu.edu.

Texas A&M AgriLife Extension Service Soil, Water, and Forage Testing Laboratory
2478 TAMU
College Station, TX 77843–2478
Telephone: 979–845–4816
Fax: 979–845–5958
Website: http://soiltesting.tamu.edu/
Email: soiltesting@ag.tamu.edu

Texas Plant & Soil Lab
5115 West Monte Cristo Rd.
Edinburg, TX 78541
Telephone: 956–383–0739
Fax: 956–383–0730
Website: http://www.texasplantandsoillab .com/
E-mail: info@tpsl.biz

Raised Bed Construction

For the edible landscape gardener, there are numerous benefits from planting in raised beds. Most edible plants prefer well-drained beds, and many will die when planted in heavy soils. Raised beds provide deep, loose soil. A greater depth of organic matter, compost, and soil amendments can be provided to plant roots by building up rather than digging in. Some deeper-rooted plants will appreciate the ground soil being amended as well as the soil in the raised bed. Another benefit is that the soils warm more quickly in the spring to get a head start on the season (soil warming can also be dependent on soil color and moisture levels). It is easier for the gardener to reach the plants, and

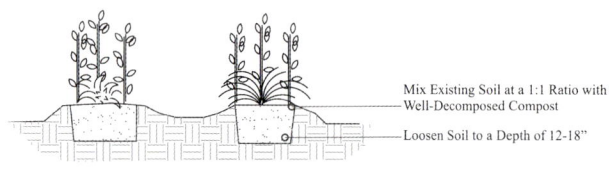

Mix Existing Soil at a 1:1 Ratio with Well-Decomposed Compost

Loosen Soil to a Depth of 12-18"

Raised Vegetable Bed Planting

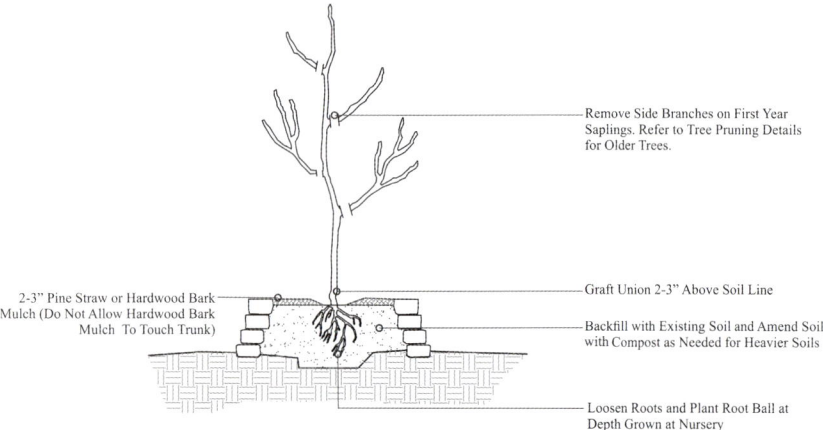

Remove Side Branches on First Year Saplings. Refer to Tree Pruning Details for Older Trees.

2-3" Pine Straw or Hardwood Bark Mulch (Do Not Allow Hardwood Bark Mulch To Touch Trunk)

Graft Union 2-3" Above Soil Line

Backfill with Existing Soil and Amend Soil with Compost as Needed for Heavier Soils

Loosen Roots and Plant Root Ball at Depth Grown at Nursery

Raised Bed Tree Planting

planning and design process. The extra effort put into design and construction of raised beds in the planning stage will pay off tremendously in the lifetime of the garden. The following are two raised bed construction methods.

Wire mesh: Cut pieces of 4- by 4-inch wire mesh into individual 4- to 8-inch strips, and curve the bands in different shapes. This is a very effective method for creating unusually shaped beds. It is important to line the wire frame with straw to keep the soil from falling through the holes.

"Lasagna" bed: Anchor 2-foot-wide wire-mesh fencing with rebar posts that have been hammered into the ground. The mesh of the wire should be large enough to allow water but not bedding material to flow through easily. Inside the frame, place a layer of cardboard on the bottom, and build layers of plant debris and vegetable scraps like compost lasagna above the cardboard. If this is done in the fall, beds should be ready (and the cardboard deteriorated) in time for spring planting.

raised beds provide a structural element for the garden that adds interest and height variability.

Construction materials and height of the raised beds will vary. Common materials used in raised bed construction are metal edging, cedar lumber, cinder block, chopped block stone, felled trees from garden clearing, wire mesh, and landscape timbers. Masonry can be mortared or "drystack." It is very important to lay out any mortared and other permanent elements as a part of the site

Fertilization

In addition to healthy soil preparation, a regular feeding program will aid plant health and vigor. The following products are recommended as a part of a regular maintenance program to keep plants producing at their optimal potential.

Fertilization: A monthly foliar application of fish emulsion and seaweed with molasses during the growing season is recommended. An additional application in the early spring is a good idea to get the life going in the soil and help plants resist damage from late freezes. A simple backpack sprayer makes larger areas easier to spray. They are not very expensive and cut the time for this chore considerably. Compost tea can be applied at the same time, but it should be applied only to the soil. There has been some indication of disease suppression if compost tea is applied as a foliar feed; however, there is some question about the safety of manure products applied to edible parts of plants. The fish emulsion and seaweed are best done as a foliar feed, as plants can absorb the nutrients through stoma (pores) on their leaves as well as through the roots. A drench of fish emulsion and seaweed poured at the roots at the time of planting helps the plants deal with transplant shock while giving them a healthy feeding.

Compost: After initial soil building, compost should be applied as a topdressing to plants. This can be done as compost is finished, being careful to avoid direct contact between the compost and the plants themselves, especially during the hotter months. Direct contact can cause the stems and lower leaves to burn and can cause fungal problems. Generally, fall is a good time to apply compost; however, there are some edible plants that should be composted at other times of the year. Check requirements for individual plants. Compost acts as an insulating blanket and feeds the roots during the winter months. Compost should not be applied at the time flowers are budding or about to bud on fruit trees, because it encourages green leafy growth and the goal at this time is to have the plant send energy to flower and fruit production.

Compost tea: Compost tea is made by steeping compost in nonchlorinated water to create a "brew" of microbes and beneficial fungi that is then applied to the soil. It is important to stir the tea daily or use an aerator to keep the tea agitated. The tea is relatively simple to make at home, or it can be purchased at local nurseries. Compost tea should be used within a few hours after it is made to ensure potency. Applications of compost tea act as a catalyst for soil microbial activity. Do not spray compost tea on edible plant parts because of potential contamination from manure in the compost.

Fish emulsion: There are two types of processes that produce fish emulsion. Hydrolyzation is a cold process that renders a product very high in proteins, amino acids, and micronutrients in a chelated form so they are immediately available to the plants. It will also go through drip systems more easily without clogging them. An emulsifying heat process produces a thicker, smellier product that can clog emitters. Both types of fish emulsion are good fertilizers, especially in combination with seaweed and molasses. Fish emulsion can be applied with a watering can and is very good for "watering in" tender seedlings. For more mature plants, apply as a foliar spray for best results.

Iron chelates and foliar spray: Iron chlorosis shows up as interveinal yellowing in new growth. It is caused by an iron deficiency and is especially common in high-alkaline soils. The soil pH can be altered somewhat with compost amendment, but it may still be necessary to apply iron chelates to the soil or foliar iron sprays. Some plants are more susceptible to this problem, and they may need to be avoided if the pH cannot be lowered sufficiently. Be cautious in the application, as the iron will stain hard surfaces.

Molasses: The sugars in molasses act as a food for microbes and have been shown to reduce the number of harmful nematodes in the soil. Other benefits include natural B vitamins and a high iron, sulfur, and micronutrient content. Molasses can be added to fish emulsion and seaweed applications to give them an extra boost. This product also comes in a dry form.

Seaweed: As a foliar feed that protects plants from stress, seaweed is high in amino acids, hormones, enzymes, and vitamins. It is an all-around excellent organic plant fertilizer.

 Irrigation

In the heat of the Texas sun, very few plants will survive without a consistent source of water. Drip irrigation is the best method for watering edible gardens for a number of reasons. The water goes straight into the soil and directly to the plant roots. In spray systems, there is quite a lot of water lost from evaporation and wind, which blows the tiny water droplets away from the intended target. Another good reason to water edible plants with drip systems is to prevent water from getting on the leaves. Soil- and waterborne fungal diseases are a real problem when growing edible plantings, and the less water on the leaves to carry these pathogens, the better. Drip irrigation also conserves water. Even the most well-designed spray system will have some overspray onto hard surfaces, so the water will be lost down the storm drain. Because drip systems are more efficient in the delivery of water to the roots of the plants, less water is used. This saves a precious natural resource as well as money on the water bill. Drip lines can be covered with mulch if desired. The concept of a drip system is really very simple, and several types of systems are available, each with certain benefits and drawbacks.

Drip system: Although all three of the water delivery systems described in this section are "drip" systems, this is the specific description for this type of irrigation system. The drip system comprises a polyethylene pipe with emitter tubing running from the pipe to the individual plants. This is a fairly simple system to construct. The feeder hose is run through the length of the bed, and a special tool is used to puncture the hose at intervals to attach the "spaghetti" tubing, which is run to the individual plants. A number of tubes can be attached at each puncture to rings of emitter tubing connections. Specific nozzles can be attached to the tubing for drip application, or spray emitters can be installed that emit tiny fans of water. A filter screen is installed at the hose bib connection to trap sediment and minerals that might clog the emitters. Filter screens should be cleaned periodically. Thick fertilizers such as fish emulsion will clog the lines, so it should not be applied in this way. Timers are available for drip systems and can be programmed like an overhead spray system.

Soaker hose: This is probably the simplest drip irrigation system. It is a hose commonly made from recycled tires that has tiny holes along the length that seep water. The hose is simply laid on the ground and can be snaked through the bed area close to the base of the plants. A couple of coils may be required around trees to supply more water. (It is still a good idea to soak trees deeply once or twice a month during the summer with a dripping hose for a few hours per tree.) Two or three lengths of hose can be supplied from one source if the pressure is sufficient. Additional lengths of soaker hose will need to be hooked up to the water supply at a different time. Having to reconnect the hose to the source on larger gardens is one drawback to the soaker hose system. Occasionally, animals will chew through the hose to reach the water source. If these animals are in the edible garden, there are probably bigger concerns than the irrigation hose.

T-tape: This is a version of the soaker-type system made from a heavy-gauge plastic. The emitters are evenly spaced, every 4, 6, 8, or 12 inches. Wall thicknesses are from 6 to 15 mils. This type of irrigation system is mostly suited to long-row planting, as the pipe does not bend well without crimping.

When attaching any drip system, it is a good idea to put a "Y" nozzle on the hose bib so the line can be used for other purposes without having to remove the drip system connection. It is also a good idea to anchor the hose in place with "U"-shaped pegs to keep it from being moved away from the plants. With any irrigation system, it is important to consider the number of emitters and the length of the run of pipe. Because of the slow water delivery, drip emitters require much less pressure (psi) to feed the system; however, there will still be a limit to the number of emitters the system can feed. Most drip emitters release ½ to 1 gallon of water per hour. A pressure regulator and backflow device are required to lessen the water flow into the tubing and prevent contaminated water from reentering the water supply. Valves can be installed along the line to turn different sections of the system on and off for longer runs of hose.

T-tape irrigation

☀ Cover Crops and Mulches

Bare soil is an open invitation for weeds and fungal disease in the garden. Planting cover crops and applying mulch can prevent this from happening.

Cover Crops

Cover crops are living mulches. They are nitrogen fixers and supply nitrogen to the soil while keeping weeds from invading the garden. Cover crops are generally planted in preparation for edible plants rather than in the same bed. Some gardeners recommend using the cover crops as living mulches around edible plantings. If this method is tried, the cover crops need to be kept low to allow good air circulation around the plants.

Fall-planted cover crops include alfalfa, Dutch white and red clovers, Elbon or cereal rye, annual rye, vetch, and peas. The clovers are naturally low, about 8 to 12 inches, and can be used effectively for living mulches. Alfalfa, ryes, vetch, and peas are taller and will be more effective as off-season plantings in nonproducing beds. Summer cover crops include black-eyed peas (cowpeas), beans, and buckwheat. Most cover crops are legumes that require an inoculation of *Rhizobium* bacteria to create a symbiotic environment that increases the plants' ability to fix nitrogen. *Rhizobium* inocu-lants are specific to the type of crop they are inoculating and are available at most garden centers and through seed catalogues.

Mulch

Mulch serves a number of purposes in the garden. It helps regulate soil temperature, keeping it cooler in the summer and warmer in the winter. Mulch protects the water in the soil from evaporating and prevents weed seeds from coming in contact with the soil, thereby preventing their germination. Mulches are generally applied from 2 to 3 inches thick. The growing season can be lengthened in winter months by pulling loose mulches (straw, hay, or pine straw) around plants when temperatures drop below freezing. Be sure to pull the mulch back away from the plant when temperatures warm.

It takes a little extra effort to go back and check the plants after applying the mulch to make sure the mulch is not covering any leaves and is not in direct contact with stems and trunks, but the effort will prevent future problems with plant loss from fungal diseases. Several types of mulches can be used in an edible garden.

Decomposed granite: This is a fine-grained

Pine straw
mulch

granite mulch that releases beneficial minerals slowly into the soil; however, it has no organic matter to break down and add to the soil's nutrients. It can be very pretty in seating areas and pathways. Decomposed granite is also effective in areas that are prone to fungal problems, as it protects the plants from the soil but does not become food for harmful fungi.

Finely shredded hardwood bark mulch: The name of this product is pretty descriptive of what it is—shredded bark from hardwood trees. This dark brown mulch is a by-product of tree-clearing projects and is probably the most commonly used mulch in landscaping projects in Texas for a few good reasons. First, the fibers compact together and the mulch will usually stay in place. Heavy-coursing water will dislodge this mulch, but that should not be a problem in a well-designed landscape. Second, this mulch will break down into food for microbial activity and help feed the life in the soil. Third, finely shredded hardwood bark mulch maintains a clean barrier between the soil and the plants. It does not tend to harbor insects and disease. Occasionally, button-type fungi or white hyphal fungi will grow in the mulch, but this fungus is not harmful to plants. It is only breaking down the mulch into soil nutrients.

Hay and straw: Gardens are often mulched with old or "spoiled" hay that has become wet and is no longer useful to the rancher. The hay can usually be obtained for little money. It is a perfectly fine mulch except that it may still have some seed heads that could germinate in the garden. It is not usually a big problem to weed the germinat-ed hay out of the garden. Straw should not have many seed heads and may be preferred for this reason. For organic gardens, it may be a problem to find hay and straw that have not been treated with chemicals. It is especially important to make sure there are no residual herbicides.

Pine straw: Pine straw is baled pine needles and makes excellent mulch for Texas gardens. The texture is perfect for matting and staying loose at the same time, allowing air circulation through the mulch while maintaining the beneficial properties of weed suppression and soil temperature regulation. Pine straw can be somewhat difficult to find in areas where pine trees are not grown. It can also be quite a bit more expensive than other types of mulch. On the plus side, it does not tend to break down as quickly as other types of mulch. Pine straw is also beneficial in providing some acidity to alkaline soils.

Pine bark: Pine bark is coarse-textured bark mulch that can have a tendency to harbor insects and float out of beds when it is used. It is not recommended for these reasons.

Other mulches: Some people use ground pecan shells, but there is some question about whether or not members of the *Carya* genus (like pecans) might emit a plant growth inhibitor. This will be good for suppressing weeds around trees, but it would not be a good idea in garden beds around more tender plants. Cocoa shells have also been used as attractive and fragrant mulches. This mulch should not be used around dogs because cocoa is toxic to canines.

PART II

Design

Bed with garlic, Swiss chard, and beets (Photo by Andrew Edmonson)

✳ Designing with Edible Plants

Merriam Webster defines *design* as "to plan and make decisions about something that is being built or created: to create the plans, drawings, etc., that show how something will be made." The action of design is more nebulous and involves the ability to imagine how a created work will evoke sensation, to be able to visualize how the final project will look before it exists. Texture, size, color, and form all combine to create interest in the garden. A kind of "inner eye" is required to foresee how these elements will play off one another, contrasting and enhancing in a way that will make an impact the individual plants will not achieve.

Even though the addition of edible plants increases a designer's palette, edible landscape design presents two unique problems. The first involves the harvest. It is important to find a balance between not wanting to harvest for fear the garden will look bare and harvesting everything so there is nothing left in the garden. This problem can be solved with the judicious use of evergreen and perennial plants to hold the structure of the garden and by harvesting only parts of the plant at a time when possible. Harvesting kale is a good example, because only the lower leaves are taken, leaving the top growth. The second problem is that many of the plants are annuals and will finish their season at about the same time. Timing plant rotation in edible garden design can be a very delicate trick. Again, the use of evergreen and perennial plants will help fill in the bones of the garden. Further filling can be accomplished by using annual color or reseeding flowers to brighten bare spots. The garden will in fact look a little bare in spots at times and sparse at others. It may be tempting to crowd plants together to create a fuller look, but the best plan is to space them according to their mature size to allow for good air circulation and sunlight penetration and to prevent the plants from having to compete for nutrients. Like a chess game, the edible garden design is consistently evolving and the designer should always be thinking of the next three steps, to be prepared to fill a bare spot or to plant in the shadow of a plant that will soon be removed. The following design guidelines will help in laying out plants for the greatest seasonal effects.

Design Elements

The first part of this design section categorizes plants according to their design uses in the contexts of texture, size, form, and color, to aid the designer in selecting the perfect plant for a particular design characteristic. As in other forms of design, the designer will usually find that only a handful of selections will suit a particular situation. The second part offers design examples that demonstrate specific design styles and the ways edible plants can be used to create the desired appearance for a particular site.

TEXTURE

The texture of a plant is commonly described in terms of being "bold" or "fine," usually dependent on the size and density of the leaves. Examples of bold-textured plants are artichokes and figs, which have large, deeply lobed leaves and make a strong statement in the landscape. Fine-textured plants generally have a light, airy appearance. Examples of fine-textured edible plants for Texas gardens are asparagus and carrots. Bold-textured plants can be used for a number of purposes: as intermediate punctuation in a formal or modern garden, a dense screen, a frame to draw the eye to a particular element, a counter to fine-textured plantings, and a stand-alone plant as an anchor or a specimen in the garden. The eye is usually drawn to these elements before the fine-textured plantings. In contrast, fine-textured plants act as a binding agent for the garden. They can be used as a soft background, as a "sea" of contrast in which the bolder-textured plants "float" or en masse as a textural element of their own. Following are examples of bold- and fine-textured edible plants:

Bold-textured plants
Artichoke
Citrus
Eggplant
Fig
Garlic
Kale, especially 'Dwarf Blue Scotch' and 'Toscano'
Leek
Lemon grass
Loquat
Okra
Savoyed cabbage varieties
Squash
Swiss chard

Fine-textured plants
Asparagus
Brassica flowers
Carrot
Coriander
Dill
Elderberry
Fennel
Mustard greens, 'Ruby Streak'

SIZE

Size of plants is an important consideration when laying out a garden, keeping in mind that taller plants will form a backdrop or punctuation. Medium-sized and shorter plants should be stairstepped down to the front of the view so that taller plants do not block them. Size categories, including shade trees, shrubs, and ground covers, are not definitive, as there are a variety of sizes within a particular type. For example, there is a very blurry line between large shrubs and small trees.

Many designers draw symbols to designate size on their plans and then assign a suitably sized plant that will fit the environmental conditions. A palette of desired plants is used to paint in the specific plants onto the garden canvas once the size has been determined. Following are examples of sizes of edible plants:

Shade tree
Italian stone pine
Pecan
Walnut

Ornamental or specimen tree
Apple
Apricot
Citrus
Jujube
Loquat
Mulberry
Nectarine
Olive
Peach
Pear
Persimmon
Plum
Pomegranate

Shrub
Bean, bush varieties
Blackberry, shrub varieties
Blueberry
Dwarf citrus
Dwarf peach
Dwarf pear
Dwarf plum
Dwarf pomegranate
Elderberry
Fig
Kumquat
Rosemary

Ground cover and border plants
Mint
Oregano
Salad burnet
Strawberry
Sweet potato
Thyme

Vines
Beans
Cucumber
Grape
Malabar spinach
Melons
Peas
Pumpkin
Squash

FORM

The form of a plant refers to its growth habit, such as columnar, arching, pyramidal, or sprawling. Many times the form of a plant is created by special pruning techniques, such as "lifting" small trees or large shrubs by pruning lower branches from the base. Topiaries and espaliers are good examples of pruning techniques that have resulted in a particular form. *Form* is the root word of *formal*, and the use of formally pruned plants is a technique to use when designing formal gardens. Select low, compact plants for a formal border for parterres, and repeat a formal geometry with columnar varieties of fruit trees, topiaries, and repetative plant patterns. Following are examples of edible plant forms:

Arching
Elderberry
Mulberry
Pomegranate

Columnar
'North Pole' columnar apple
'Scarlet Sentinel' columnar apple

Espalier
Apple
Pear
Plum
Pomegranate

Screens
Bay tree
Blueberry
Citrus
Corn
Elderberry
Fig
Jujube
Loquat
Pomegranate

Topiary
Bay tree
Citrus
Rosemary

Umbrella
Italian stone pine
Loquat

COLOR

Color is perhaps the greatest tool a landscape designer can employ to add interest to the garden. Foliage, fruit, and flowers can all be interesting at varying times of the year, and certain colors play well together. Colors that are opposite one another on the color wheel work well in combination. Examples include red and green, purple and yellow, and orange and blue. Color can also be used to induce a particular experience in the garden. For example, a cooling effect can be achieved with blue, white, and light yellow. A vibrant effect can be achieved with the hotter colors: red, orange, and bright yellow. Varying hues of purple, gray, and silver and variegated foliage can be used in combination with either palette as a binding agent. Green is the most common color for foliage and will generally be the structure for the rest of the garden. In areas where perennials go dormant, it is valuable to have a good skeleton of evergreen structure to hold the garden together during the winter months. It is important to combine color seasonally, at the height of interest. Another method that works well for color combinations is to repeat varying hues of the same color, not as a single palette necessarily but as a theme to bind the whole. Following are examples of edible plants by color:

Blue
Rosemary, flower

Evergreen
Bay tree
Citrus
Italian stone pine
Loquat
Olive
Oregano
Rosemary
Thyme

Orange
Cucumber, 'Hmong Red,' fruit
Eggplant, 'Goyu Kumbo,' fruit
Kumquat (some varieties), fruit
Loquat (some varieties), fruit

Orange, fruit
Pepper, 'Doe Hill Golden Bell,' 'Habanero,' fruit
Persimmon, fall foliage and fruit
Pomegranate, flower
Pumpkin (some varieties), fruit
Swiss chard, 'Oriole Orange,' stalks

Pink
Chives, flower
Elderberry, 'Black Beauty,' 'Black Lace,' flower
Peach, flower

Purple
Artichoke, flower
Asparagus, 'Purple Passion,' spears
Asparagus bean, 'Red Noodle,' pod
Bean, 'Blue Coco,' 'Dean's Purple Pole,'
    'Purple-Podded Pole,' 'Royalty Purple Pod,'
    pod
Broccoli, 'Purple Sprouting,' fruit
Brussels sprouts, 'Falstaff,' 'Red Rubine,' foliage
    and fruit
Cabbage, red varieties, foliage
Cauliflower, 'Purple of Sicily,' 'Violetta Italia,'
    fruit
Eggplant (some varieties), fruit
Elderberry, 'Black Beauty,' 'Black Lace,' foliage
Fig, "Violet de Bordeaux,' fruit
Kohlrabi, 'Purple Vienna,' foliage and fruit
Lemon, 'Eureka,' purple tinges on new growth
    and flowers
Mustard greens, 'Osaka Purple,' foliage
Peas, 'Blue Podded,' pod
Pepper, 'Black Pearl,' foliage and ripening fruit

Plum, 'Methley,' 'Stanley,' fruit

Tomatillo, 'De Milpa,' 'Purple Coban,' fruit

Red

Amaranth (some varieties), seed heads and foliage

Apples (some varieties), fruit

Bean, 'Scarlet Runner' (orange-red), pod

Bee balm, flower

Blackberry, fruit

Blueberry, fall foliage

Grape, fall foliage

Kale, 'Russian Red,' foliage

Lettuce, 'Cimarron Red,' 'Danyelle,' 'Lollo Rosso,' 'Merlot,' 'New Red Fire,' 'Outstanding,' 'Red Deer Tongue,' 'Red Sails,' 'Red Velvet,' 'Rouge d'Hiver,' foliage

Loquat, 'Early Red,' fruit

Mulberry (some varieties), fruit

Mustard greens, 'Red Giant,' 'Ruby Streak,' foliage

Okra, 'Alabama Red,' 'Bowling Red,' 'Burgundy,' 'Hill Country Heirloom Red,' leaf veins, stems, and pods

Pepper, 'Aci Sivri,' 'Cayenne,' 'Cubanelle,' 'Papri Sweet,' 'Sweet Red Cherry,' 'Thai Hot,' fruit

Persimmon, fall foliage

Plum, fruit; 'Allred,' leaves and fruit

Pomegranate, fruit

Spinach, 'Malabar,' leaf veins

Strawberry, fruit

Swiss chard, 'Ruby Red,' foliage

Tomato, fruit

Silver/Gray

Artichoke, foliage

Collard greens, 'Georgia Southern,' foliage

Kale, 'Dwarf Blue Scotch,' 'Toscano,' foliage

Leek, foliage

Olive, foliage

Sage, foliage

Thyme, foliage

White

Anise, flower

Apple, flower

Blackberry, flower

Chamomile, flower

Chervil, flower

Cilantro, flower

Citrus, flower

Cucumber, 'White Wonder,' fruit

Dewberry, flower

Eggplant (some varieties), fruit

Garlic chives, flower

Loquat, flower

Pear, flower

Raspberry, flower

Swiss chard, 'Lucullus,' 'Schnittmangold Gelb,' stalks

Yellow

Amaranth (some varieties), seed heads

Apple (some varieties), fruit

Brassicaceae flowers

Cucumber, flower; 'Chinese Yellow,' 'Lemon,' fruit

Dill, flower

Elderberry, 'Aurea,' foliage; 'Goldbeere,' fruit
Fennel, flower
Grape, fall foliage
Jerusalem artichoke, flower
Lemon, fruit
Lettuce, 'Lollo Biondo,' foliage
Loquat (some varieties), fruit
Mexican mint marigold, flower
Pepper, 'Aji Amarillo,' 'Banana,' fruit
Plum, 'Wickson,' fruit
Squash (some varieties), fruit, flowers
Variegated thyme, foliage

Traditional Gardens: Parterre and Potager

Both the parterre and potager garden design styles reached their height during the Renaissance. In the parterre style the garden beds are divided into geometric patterns, both the beds and the plantings within the beds. The potager is a kitchen garden, traditionally the garden that was closest to the house for harvesting daily herbs, vegetables, and fruits, while the fields would be located farther from the house. A potager may be designed in the geometric style of the parterre, so a potager may actually be both. While kitchen gardens have existed since humans first began cultivating edible plants, the formalized parterre garden style is more recent. Originating in Italy, the parterre style was brought to France, transforming French gardens and creating ornate garden tapestries of color. The most well-known gardens of this style are those of the French aristocracy. The gardens

Potager du Roi, Palace de Versailles, France (Photo by Sumiyo Ida)

at Versailles, Vaux-le-Vicomte, and the Palais des Tuileries in France are magnificent examples of this garden style. Parterre and potager gardens do not need to be as grand as the Potager du Roi. These styles adapt very well in size and scheme to fit typical home landscapes.

PARTERRE GARDEN

The parterre is a formal landscape design style, with repeating geometric patterns that often include formally clipped hedges and borders. Beds are commonly laid out around a central focal point with strong axial and terminus elements. Garden ornaments, fountains, pergolas, trellises, and sculpture are typically used to create these features.

Parterre at Lake Austin Spa Resort

Alignment and ornamentation: In the parterre, plants are used as architectural elements and are often sheered into geometric shapes, such as cones, boxes, and spirals. Ornamental trees can be used to form grids in the garden or planted to form allees (rows of evenly spaced trees that border and frame drives or wide walkways). Shade trees may be planted in this fashion along long, wide driveways. Along more narrow pathways, fruit trees can be used to construct gorgeous flowering arches for dramatic entrances and enchanting walkways.

Pathways: Geometrically aligned pathways are used to separate the beds and align the landscape. These pathways may be lawn, gravel, mulch, or paved surfaces and should be aligned with architectural structures. When using lawn as a choice for pathways, it is important to have a substantial bed edge of stone or heavy-gauge steel to prevent the invasion of lawn grass into the beds. It is also important to make sure paths are wide enough to sustain lawn growth, at least 3 to 4 feet.

Border: The beds are usually bordered with a common border plant. English boxwood and dwarf boxwood (*Buxus* spp.) are often used for this purpose. Select dwarf varieties for larger bed spaces (3 to 4 feet tall) and very dwarf varieties for smaller spaces (1 foot tall). Green creeping germander (*Teucrium chamaedrys*, not *T. cossonii*) is also a good choice. It is lower growing than dwarf boxwood cultivars and can be shaped into a tight border. For a fragrant border, thyme (*Thymus* spp.) and garden chives (*Allium schoenoprasum*) are good choices. Sometimes, these border plants are used inside the beds, dividing them into intricate patterns. This is called a "knot garden" and is particularly pretty planted in various types of herbs.

Plant location: Plants are laid out within the beds in a symmetrical pattern, layering the plants from shortest to tallest. In beds that are surrounded on all sides by pathways, the tallest plants should be located in the center. For beds located on the garden perimeter, taller plants are usually placed in the back. The geometry is repeated in the garden in similarly shaped beds to provide a sense of continuity and flow to the overall design. Dwarf fruit trees, especially tender citrus varieties, can be planted in large pots and used as accents instead of in-ground plantings.

Suggested edible plants: All of the plants in this book are suitable for a parterre garden. It is more important to consider repetitive patterns and how the plants will work together as a whole. A well-kept common border with repetitive upright elements will allow for more informal plantings within the border. It is not necessary that the filler plants be planted in repeating patterns.

- Fruit trees: Dwarf and semi-dwarf varieties for interior garden beds and standard size for perimeter plantings and drive allees in larger gardens; citrus in pots or planted in the ground in warmer regions. Olive, bay, and myrtle trees can be sheered into topiaries for accent.
- Evergreen herbs: To hold the garden structure during winter months. Rosemary, sage, oregano, garden chives, marjoram, and thyme are good choices.
- Grapevines: Good for trellising on arbors and pergolas.

POTAGER GARDEN

The potager (kitchen garden) has traditionally been a garden located within easy access of the kitchen for harvesting edible plants for everyday use, while grain and forage crops were grown in fields farther from the house and less accessible. The kitchen garden typically contains a variety of edible plants that a family uses on a daily basis. This type of design might encompass an entire yard or may be a simple planting of a few herbs and vegetables within easy access of the kitchen door.

Layout: Potagers can be laid out in formal geometric shapes or loosely structured with meandering pathways in a cottage garden design. The garden is usually enclosed within some type of fencing or wall and has pathways that allow the gardener to reach the plants comfortably.

Suggested edible plants: Traditionally, the kitchen garden included a variety of plants, including culinary and medicinal herbs, flowers, and vegetables, with the various colors and textures woven into a beautiful collage. Beds may be bordered with a common border plant as in a parterre design, although this is not a requirement, and potagers are usually less formal. The most important consideration in plant selection for a potager is what the most commonly used herbs and vegetables will be.

PARTERRE GARDEN PLAN: THE GARDENS AT LAKE AUSTIN SPA RESORT

Lake Austin Spa Resort is a luxury spa with forty guest rooms located on Lake Austin in Austin, Texas. Aster Café, a full-service kitchen that serves breakfast, lunch, and dinner, is located at the resort and supplied by an extensive edible garden. The garden, designed by renowned landscape architect Eleanor McKinney, is beautifully tended by gardener Trisha Shirey and utilized by French chef Stephane Beaucamp to create healthful, elegant menus based on seasonal availability.

Trisha has been working at Lake Austin Spa Resort for twenty-nine years and has a strong

Potager garden (Photo by Rebecca Nickols)

Trisha Shirey, director of flora and fauna, and Chef Beaucamp, head chef, at Lake Austin Spa Resort (Photo by Andrew Edmonson)

belief in a garden that is both beautiful and useful. She began working on the edible garden on the second day of her career at the resort, building the beds over the years with compost that is made on-site from kitchen scraps and lawn trimmings as well as duckweed collected from the lake, dried, and worked into the mix. The unusually large and lush condition of the plants at Lake Austin Spa Resort is a testament to her experienced care and dedication to natural processes. Trisha experiments with unique varieties to offer otherwise hard-to-find flavors through the seasons. Guests

at the garden are often inspired to create gardens of their own after seeing the beauty of the edible garden that Trisha has built.

Chef Stephane has come to Texas from France via California. Since he had lived in areas with different climate zones and growing conditions, he had to learn the possibilities and limitations of Austin's growing conditions. Working closely with Trisha, Chef Stephane creates menus that are imagined and altered to create uniquely Texas cuisine. Chef Stephane's cooks come out with him to the garden to explore what is in season, tasting and smelling fresh herbs and vegetables and thinking about how the flavors might be combined to create daily specials. Fresh herbs, seasonal fruits and vegetables, and flowers for garnish are harvested daily and supply 25 to 30 percent of the kitchen's produce. Trisha and her gardeners harvest the produce for the kitchen to ensure that the form and integrity of the plants are preserved, both to ensure future harvests and to maintain the beauty of the garden.

Although the design has a gentle subtleness, much thought and effort have gone into creating the natural-looking gardens. The intuitive understanding of color, texture, and form within the context of a working garden is truly inspiring. Strolling through the garden is a sensual delight of color, smell, and taste within the design of a traditional European parterre. Eleanor McKinney uses a geometric layout, repeating strong architectural elements in the garden. Much of the inspiration for her designs comes from her travels in Europe. With gravel pathways and an herbaceous

border, the kitchen garden at Rousham House in Oxfordshire, England, is particularly influential in the design of the gardens at the Lake Austin Spa Resort.

In the initial site analysis, McKinney realized that the traffic circulation and site references would need to be addressed. During the carving of the softer design from the existing site, the vehicular entrance to the gardens was moved southward to allow more garden space between the drive aisle and the existing cottages. A slope from the cottages to the drive also needed to be addressed by replacing landscape timbers with stone terraces.

Eleanor McKinney at the Gardens at Lake Austin Spa Resort (Photo by Andrew Edmonson)

(below) Chefs harvesting produce in parterre garden at Lake Austin Spa Resort

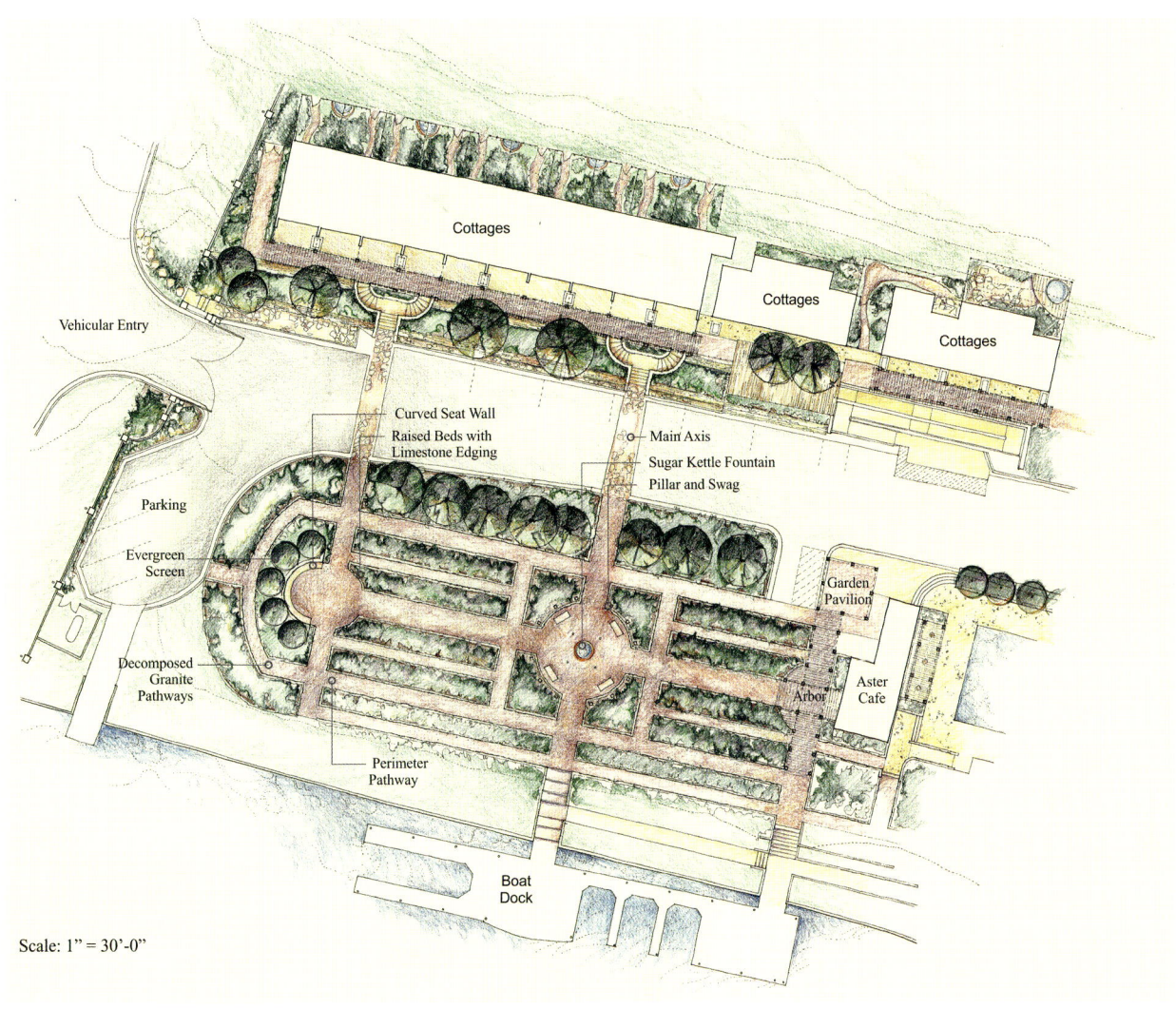

Vehicular Entry

Parking

Evergreen
Screen

Decomposed
Granite
Pathways

Perimeter
Pathway

Cottages

Cottages

Cottages

Curved Seat Wall
Raised Beds with
Limestone Edging

Main Axis
Sugar Kettle Fountain
Pillar and Swag

Garden
Pavilion

Arbor

Aster
Cafe

Boat
Dock

Scale: 1" = 30'-0"

The gardens at Lake Austin Spa Resort by Eleanor McKinney, RLA
(Courtesy of McKinney Landscape Architecture, Inc.)

The curved staircases from the cottages align with the entrances to the gardens to emphasize axial lines and views of the lake beyond. A parking lot on the west side of the garden is concealed with a curved seat wall and tall evergreen screening plants, which give the garden a more enclosed feeling. The main aisle connects the seat wall with the restaurant that lies on the opposite axis of the garden, which is accentuated and shaded by a trellis. The central focus of the garden is a sugar kettle fountain surrounded by a traditional pillar and swag created out of tapered wooden pillars to match the trellis. The raised beds are laid out symmetrically from the center with an arc and tangent stone edging. Evergreen plantings are designed into the garden to provide year-round structural form, leaving spaces for annual plantings.

Children's Garden

Children are born explorers, and the natural world is their great magical mystery. Studies in the last decade affirm that children prefer to play in and explore the natural environment rather than constructed environments that do little to fuel their imaginations. The natural environment provides learning experiences not found in the classroom. Low branches of trees become ladders, forts, and castles. A mud hole and a stick become great explorations of density and viscosity. The mysteries of a flower or the methodical movements of an inchworm delight and intrigue a child's imagination. Children can spend hours learning and absorbing all of the wonders of nature found in the garden.

Abbey with a yellow crookneck squash in the garden (Photo by Tom Cronk)

The edible garden is a wonderful place to introduce children to the natural world. The mystery of life held in a seed that grows to become a beautiful plant is a delight to children. Here they learn a love and respect for nature and an appreciation of where their food comes from. Children might even be more likely to eat their vegetables if they helped in the cultivation and harvest.

There are so many lessons in a garden, both abstract and academic. Math and science are the obvious academic connections; however, art, literature, and history have their place as well. The one key to guiding children's natural hunger for knowledge is to never make it seem like work. Inherently, a garden is a lot of work, and children should be given responsibility for specific plants and chores, but the goal (especially for young children) is to learn about the nature of garden plants. In time, the practical gardening concepts, such as methods of cultivation and ways to increase yield, should be introduced as children seek to apply the lessons they have previously observed. Middle and high school years are a good time to get down to the nuts and bolts of food production. It is sufficient for younger children just to feed their natural desires to explore their environment and to guide them in the discovery of what does (and does not) work in the garden.

Layout: A children's garden should have clearly delineated pathways and planting areas to prevent seedlings from being trampled. It helps to put plant markers where seeds are planted. These can be as simple as seed packets glued to Popsicle sticks. The packet will probably not deteriorate before the plants are up and identifiable.

Time commitment: Time is usually a precious commodity for those who care for young children, and a small successful garden is a lot more fun for everyone concerned when time for maintenance is limited.

Bean and pea tepees: A simple construction can be made with long bamboo poles stuck in the ground and tied at the top with jute or string, allowing vining plants' tendrils to encircle the poles. If the poles are at least 6 feet out of the ground, a little room can be made inside the tepee for a natural hiding spot.

Pear harvest, Austin, Texas

Natural mazes or rooms: Children enjoy a sense of enclosure and fun "rooms," or even whole mazes can be created with taller plants such as cornstalks or sunflowers. For more naturalistic spaces, leave branches of trees or large shrubs low to the ground and prune out inner branches.

Water: Water features can be wonderful elements for play and exploration. It is a good idea to provide water stations for children to fill up containers to water their plants. A simple wooden stand, at least 2 feet in height, with a cutout for a dish tub will work. Shelves or ledges for small watering cans should be a part of the construction. It is important to consider the age of the children when planning a water feature. Well-designed water tables are low enough for children to dip watering containers in but tall enough to prevent them from falling into the basin.

Wildlife habitat: Although most gardeners try to avoid attracting unwanted scavengers to the edible garden, a wildlife habitat is a nice feature to include in a children's garden. Butterfly gardens are very popular with children, and nectar and habitat plants can act as decoys for caterpillars and moths that might otherwise feed on the tomatoes, and they also attract pollinators. Birdbaths, bat houses, and bird feeders make nice features for enjoying wildlife.

Whimsical features: Children enjoy seeing interactions in nature. The wind playing with a pinwheel or a wind chime can be mesmerizing. The shadow cast by a sundial can teach older children about the earth's rotation. Water pouring from one level to another in a terraced fountain is delightful for gardeners of all ages. Again, be cautious about depth of water in basins around young children.

Compost heaps and cold frames: Children will enjoy the experience of "recycling" kitchen scraps and leaves in a compost heap and seeing the natural transformation into usable compost. Cold frames can be constructed for starting "sets" to put out in the garden at the appropriate time.

Cooking with harvested plants: Children enjoy applied knowledge, and learning how to prepare the food they have grown is rewarding and useful, allowing them a completed vision from seed to table. Children will also be more likely to eat vegetables if they have had a hand in growing and preparing them.

Suggested edible plants: It is important to consider the age and safety of the children when planning children's gardens. If young children are allowed to eat plants directly from the garden, make sure there is nothing poisonous in the yard. Exploration does not always stop at the garden gate. Hot peppers should be avoided for the same reason. Children enjoy being involved in the planning process. Even young children can be involved in selecting the plants they want to grow (and ultimately eat.) The following plants can be a good place to start:

Asparagus
Baby carrots
Basil
Blackberries, thornless variety
Cherry tomatoes
Corn: Popcorn and colorful ornamental varieties can be fun (stagger plantings of sweet and traditional varieties to prevent cross-pollination).
Edible-podded peas
Fruit trees: Leave boughs low to ground for natural hideouts (mulberries and figs work especially well).
Garlic
Lemon balm

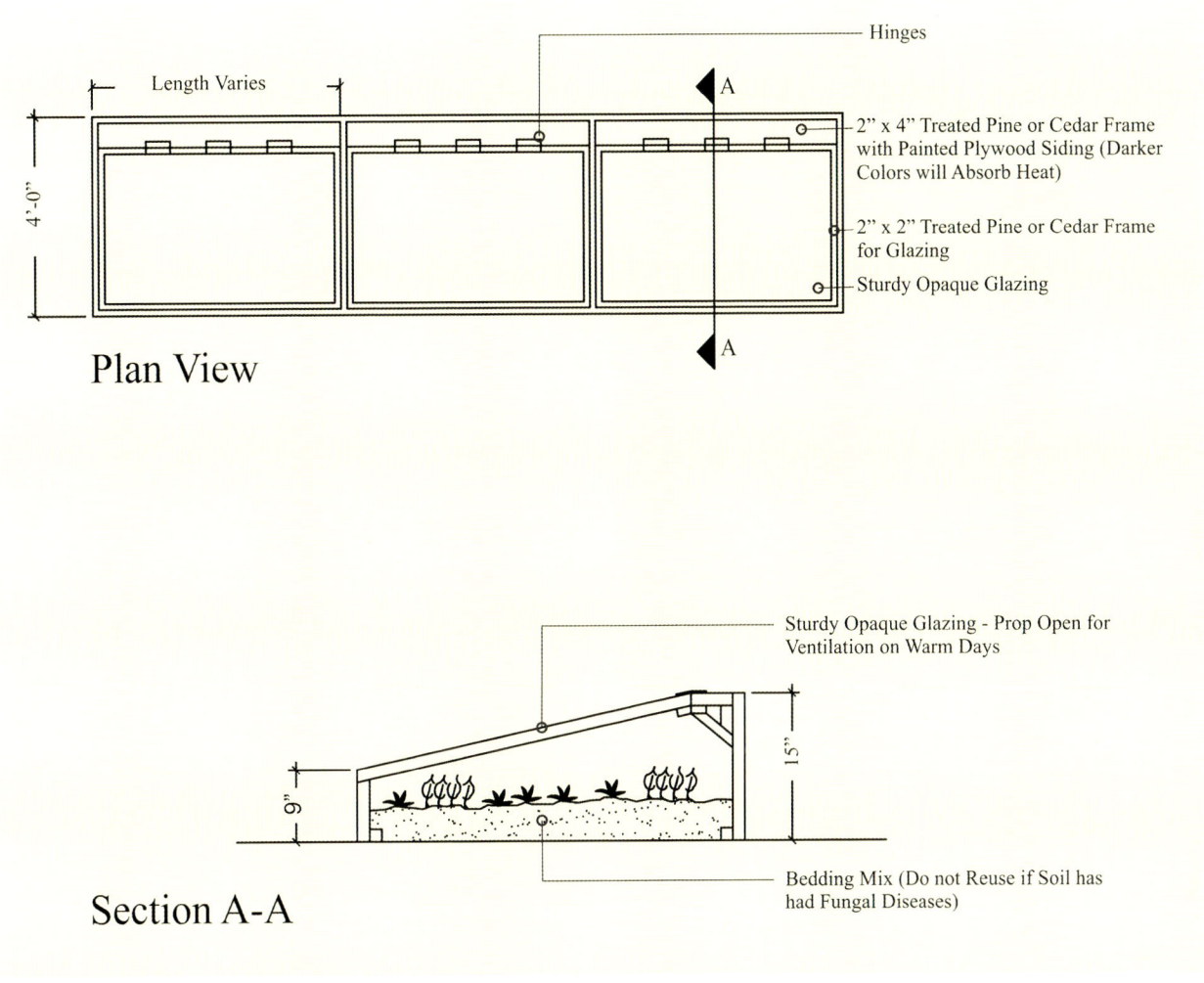

Plan View

Length Varies

4'-0"

Hinges

2" x 4" Treated Pine or Cedar Frame with Painted Plywood Siding (Darker Colors will Absorb Heat)

2" x 2" Treated Pine or Cedar Frame for Glazing

Sturdy Opaque Glazing

A

Section A-A

Sturdy Opaque Glazing - Prop Open for Ventilation on Warm Days

15"

9"

Bedding Mix (Do not Reuse if Soil has had Fungal Diseases)

Cold frame construction (NTS)

Miniature pumpkins and ornamental gourds
Mint
Nasturtiums
Pineapple sage
Roses, thornless or nearly thornless

Scarlet runner or other types of climbing beans
Spinach
Strawberries
Sunflowers
Violets

Mina and the bean stalk (Photo by Bradley Leeper)

CHILDREN'S GARDEN PLAN: OLIVE TREE LEARNING CENTER

The Olive Tree Learning Center is an early childhood learning center in Austin, Texas which is able to accommodate about thirty children ranging between eighteen months and five years in age. The school's curriculum includes growing vegetables from seed and small container plants for their garden, which the school's director and owners decided to expand in 2009. The decision was made to plant the entire site in edible plants, including fruit trees, herbs, vegetables, and bramble fruits. As many of the existing nonedible plants as possible on the site were kept, although some needed to be relocated.

The initial meeting between the director, owners, and designer included a discussion about specific changes to the site regarding drainage issues and playground adjustments. The schoolchildren were consulted about what vegetables they would like to grow. Specific design considerations included a poorly drained concrete area where the children ride their tricycles. This drainage was altered to prevent water and mud from collecting on the concrete, and an additional paved area was added to complete a circular path around the yard. The remainder of the site was filled with edible planting beds or pathways. Rain barrels were added to reduce runoff and to harvest rainwater for the children to use to water the plants.

The site already had some vegetable garden beds, bordered by landscape timbers. These beds

Planting day at Olive Tree Learning Center

'Zepherine Drouhin' Rose on
Cedar Lattice

Climbing Boulders

Existing Pecan

4 - 'Collier' Mulberry

Pea Gravel

Paved Area for Tricycles

Compost Bins

Playscape

Swing Set

12 - "Seascape'
Strawberry

11 - Wood Violet

Pea Gravel

Bamboo Tepees for
Trellising Beans and Peas

Pea Gravel

Wash Basin
Flagstone Walk, Typ.

Giant Sunflowers from
Seed, Typ.

2 - Pineapple Guava

13 - Mint

17 - 'Purple Passion'
Asparagus

Wood
Car

Office

2 - 'Violet de
Bordeaux' Fig

Fenced Vegetable Beds
with Pea Gravel Walkways

'Arbequina' Olive

15 - 'Sweet Charlie'
Strawberry

Rigid Metal Mesh Panels
on Cedar Posts for
Vertical Trellising

Wood
Deck

Wood
Deck

15 - 'Sweet Charlie'
Strawberry

Bird Bath

Wood Ramp

3 - Mex. Mint
Marigold

17 - 'Seascape' Strawberry

Compacted Decomposed
Granite Walkway

3 - Bee Balm

9 - Lemon Thyme

7 - 'Livin' Easy' Rose

6 - Rainbow Chard

Exist.
Red Oak

3 - Artichoke

Transplanted Shrubs
Border with 11 Oregano

3 - 'Big Jim' Loquat

Handcrafted Pavers

Existing
Pecan

Wood
Deck

Wood Ramp

Learning Center

Scale: 1" = 10'-0"

Garden design for Olive Tree Learning Center

were enhanced with the addition of pathways and a 4- by 4-inch metal mesh fence with cedar posts. This fence also served as an upright element for vining plants and included "windows" at eye level for small children, giving the fenced area the feel of an open-air playhouse. Handmade stepping-stones used to construct a walkway were personalized with hand impressions and found objects to make them unique to each of the students.

The existing play equipment remains in its current location, with additional fruit trees, edible perennials, and herbs planted on the periphery.

Little helpers making the work more like play

Mulberries and figs were designed as periphery plantings with their branches to be left close to the ground to create canopy "hideouts." The design also includes natural play areas for the children to explore in unconstructed play. Additional features include compost bins, cold frames, and plant watering stations.

Planting day at Olive Tree Learning Center was set for a Saturday in March, and many of the children and parents brought plants and assisted in planting their garden, helping them share in the ownership of the project.

Residential Garden

Residential gardens are unique to the individual property, architectural style, and preferences of the residents. The designer should tailor the layout to suit the needs and desires of those who will be living in, eating from, and tending the garden, while working within the limitations of the site and advancing the strengths. For example, a low area in the yard may be designed as a dry creek that meanders through gently curving raised beds. Time spent in site analysis and evaluation will be rewarded in the final project. Refer to part 1 of this book for more specific information regarding site evaluation and considerations in the planning process.

Styles and preferences: The architectural style of the house should be complemented by the landscaping style. Major architectural elements and patterns should be repeated in the path and bed layout for more formal gardens. A less formal cottage garden, with unceremonious pathways, might be more appropriate for a 1930s bungalow. Personal preferences and tastes of those who will be enjoying the garden space will need to be considered. Are play areas needed for small children and dogs? Should views be screened from overly inquisitive neighbors? How often will the garden be used for entertaining, and how many people will need to be accommodated? Will the garden be used for dinner parties, intimate gatherings, games, or simply personal reflection? Are there any allergy considerations? Will people who are elderly or have disabilities need special paving

Residential garden designed by Fritz Haeg

treatments and gently sloping inclines? All of these questions, and more, will need to be answered when designing for a particular family and worked into the site design.

Existing elements: Design begins with the "knowns" and moves on from there, determining what should stay and what should go. Most often, the structures and utilities will be permanent. Trees, walkways, driveways, and fences are semi-permanent and should be evaluated based on the suitability to the desired results, as well as the expense and difficulty of replacement. For example, it may not be a difficult decision to remove a sickly 'Bradford' pear that does not produce edible fruit, while the decision to remove a large shade tree that is casting too much shade on valuable growing space requires much more deliberation. In the initial enthusiasm to plant the entire site in edible plantings, the designer may be blinded to the existing plantings that provide cooling shade and a foundational planting that will leave the yard looking quite bare if they are removed. Work with the existing elements as much as possible when evaluating a site, but do not be afraid to remove something that just does not fit.

Site-specific liabilities and assets: Each residential site is unique, including strengths and weaknesses that need to be evaluated. Municipal and subdivision landscaping regulations may actually restrict the location of vegetable gardens. Drainage patterns are also a consideration. Very few edible plants will grow well in soggy conditions, so it is important to observe where runoff and pooling occur in the landscape and evaluate how water can be moved through the property without impacting plantings. Each site will have views, some to be enhanced and others to be screened. Shade patterns are another consideration, as well as circulation patterns through the garden. Spend time in various areas of the proposed garden space and within the residence looking out to the landscape to aid in evaluation before committing the design to paper.

Suggested edible plants: All of the plants in this book can be used for residential garden design in varied applications. While available space, climate, and garden style will influence plant choices, the most important consideration will be what the residents like to eat and the frequency these foods are included in meal planning. Experimentation with different species will guide future plantings.

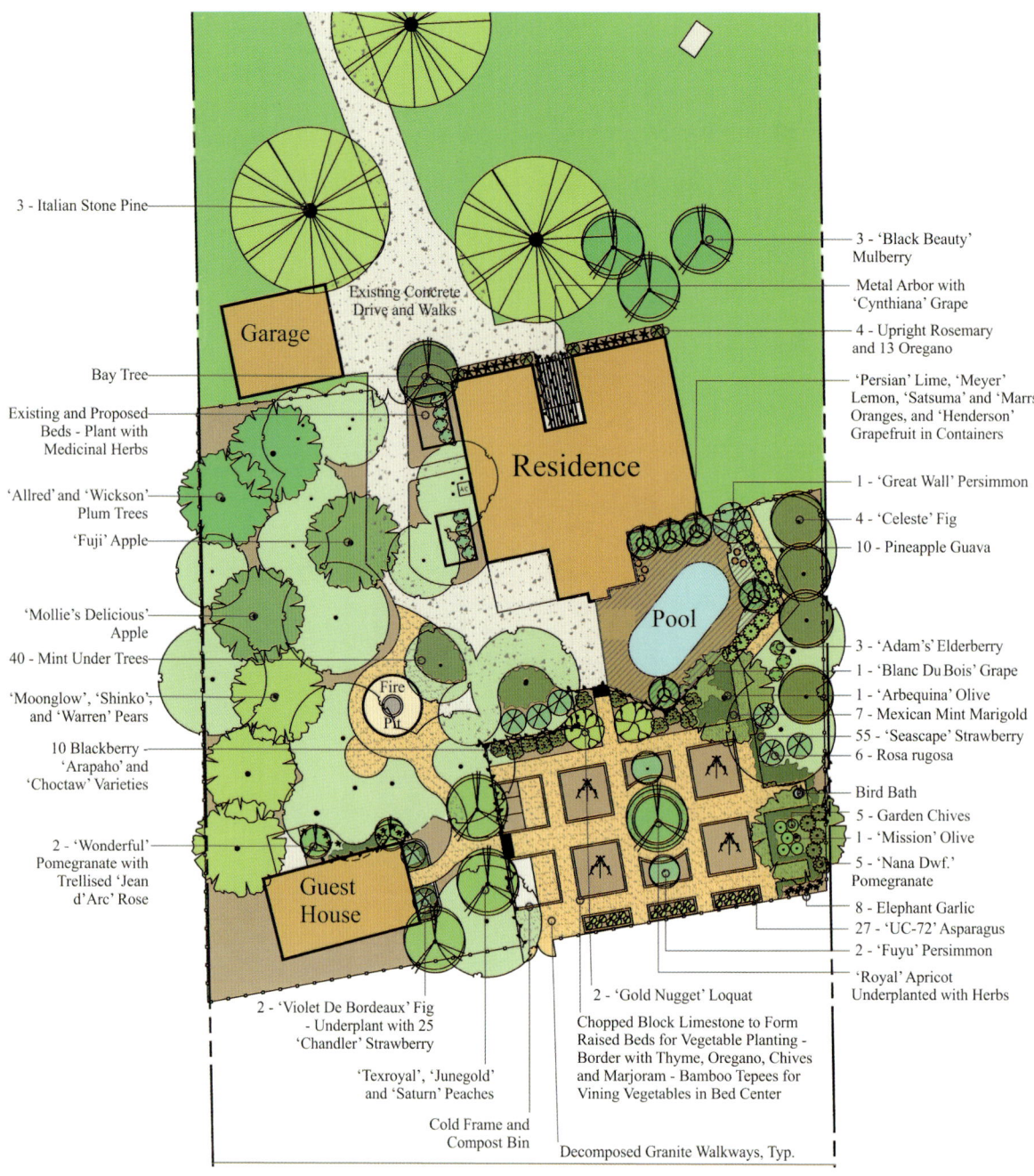

3 - Italian Stone Pine

3 - 'Black Beauty' Mulberry

Metal Arbor with 'Cynthiana' Grape

Existing Concrete Drive and Walks

4 - Upright Rosemary and 13 Oregano

Garage

Bay Tree

'Persian' Lime, 'Meyer' Lemon, 'Satsuma' and 'Marrs' Oranges, and 'Henderson' Grapefruit in Containers

Existing and Proposed Beds - Plant with Medicinal Herbs

Residence

1 - 'Great Wall' Persimmon

'Allred' and 'Wickson' Plum Trees

4 - 'Celeste' Fig

10 - Pineapple Guava

'Fuji' Apple

'Mollie's Delicious' Apple

Pool

40 - Mint Under Trees

3 - 'Adam's' Elderberry

1 - 'Blanc Du Bois' Grape

'Moonglow', 'Shinko', and 'Warren' Pears

Fire Pit

1 - 'Arbequina' Olive

7 - Mexican Mint Marigold

55 - 'Seascape' Strawberry

10 Blackberry 'Arapaho' and 'Choctaw' Varieties

6 - Rosa rugosa

Bird Bath

5 - Garden Chives

1 - 'Mission' Olive

2 - 'Wonderful' Pomegranate with Trellised 'Jean d'Arc' Rose

5 - 'Nana Dwf.' Pomegranate

8 - Elephant Garlic

27 - 'UC-72' Asparagus

Guest House

2 - 'Fuyu' Persimmon

'Royal' Apricot Underplanted with Herbs

2 - 'Violet De Bordeaux' Fig - Underplant with 25 'Chandler' Strawberry

2 - 'Gold Nugget' Loquat

Chopped Block Limestone to Form Raised Beds for Vegetable Planting - Border with Thyme, Oregano, Chives and Marjoram - Bamboo Tepees for Vining Vegetables in Bed Center

'Texroyal', 'Junegold' and 'Saturn' Peaches

Cold Frame and Compost Bin

Decomposed Granite Walkways, Typ.

Scale: 1" = 30'-0"

Starr residence garden design

RESIDENTIAL GARDEN PLAN: STARR RESIDENCE

Michael and Darlene Starr live in Manchaca, Texas, in a semi-rural area with plenty of local wildlife. Darlene is an herbalist, and she and Michael have envisioned an edible landscape that incorporates a variety of edible plants and medicinal herbs. This landscape requires a fence at least 8 feet tall to protect the plants from marauding herds of deer that live in their neighborhood.

Pathways connect the various elements of the landscape, including a guest cottage and a deck with an aboveground pool. Citrus trees are located on the pool deck in pots on wheeled platforms, which are designed to be easily moved into the garage in the event of freezing weather. The main garden is a fenced parterre, and the area outside the parterre is kept in a more naturalistic style.

In order to conserve valuable growing room in the parterre garden, they located most of the fruit trees outside the garden fence on the perimeter of the backyard, as there are quite a few shade trees in the interior. Shaded areas under the trees are utilized for entertainment, incorporating seating with a firepit, as well as a location for them to relax in their hammock after long days in the garden.

Rustic garden gates invite visitors to stroll through the garden on gravel pathways that separate the raised beds. The beds are designed to be harvested from all sides, reducing the impact of traffic within the planted area. Permanent planting beds for asparagus, blackberries, elderberries, and perennial herbs are located on the perimeter of the parterre, leaving the interior beds for rotations of annual plantings.

Commercial Garden

Restaurants, resorts, and learning centers are all examples of commercial applications for edible landscapes. Edible landscapes provide a unique experience for visitors and the allure of knowing that such care and quality are involved in menu ingredients. Food cultivation and preparation become integrated in the dining experience. Quality control, control of cost of produce, and increased variety are other benefits of edible landscapes in commercial operations.

Chef Beaucamp harvesting fresh vegetables and herbs for the kitchen, Lake Austin Spa Resort

City regulations and code: Commonly, commercial landscape design is submitted for review to city planning and building review departments, and most municipalities require drawings to be sealed by a registered landscape architect (RLA). Specific considerations include restrictive plant lists, caliper-inch requirements on trees, and watering restrictions. Large-caliper fruit trees may be difficult to locate, and sources should be found before specifying these trees to meet city requirements. Most cities allow for an appeals process and may be willing to make exceptions if edible plantings are something the city would like to encourage.

The garden experience: A well-tended garden draws visitors outside to explore the garden space, seeing, smelling, and even tasting the fresh produce. Seating can be placed within the garden for an enhanced outside dining experience. Regularly scheduled garden tours and events in the garden, such as Mother's Day tea parties and Easter brunch, can broaden the garden experience for customers.

Maintenance: It is very important that there is a qualified person to keep the garden in shape and change plants out seasonally. An unkempt garden is an eyesore and might actually discourage customers from visiting.

Suggested edible plants: All the plants in this book are suitable for commercial plantings. Menus will be shaped by seasonal availability, and plantings will be determined by the chef's desires. It is important for the chef and the gardener to work closely, planning which plants will

be grown during the seasonal planning phase. Perennial herbs and vegetables as well as fruit tree choices will be semi-permanent, requiring much consideration about varietal choices and planting quantities at the beginning of the garden design. Fresh herbs can be a high-priced item, but they are relatively easy to grow, making them a good choice for restaurant gardens. Some produce is not commercially available in most markets. Loquats are a good example, because they do not ship or store well. Unusual varieties that are not commonly available from restaurant suppliers can be grown, adding exotic flavors and colors to specialty menu offerings.

COMMERCIAL GARDEN PLAN: THE FAIRMONT DALLAS

The Fairmont Dallas is an upscale hotel located in the heart of downtown Dallas, a surprising place for a kitchen garden. The 3,000–square-foot garden is located on the fourth-floor rooftop and provides a portion of the vegetables and most of the herbs for the hotel's four-star Pyramid Restaurant and Bar. Romantic dinners in the garden are offered for small parties.

Executive Chef Andre Natera adjusts the menu as vegetables come into season. For example, the restaurant offers specialty dishes made with strawberries in the spring and tomatoes in the summer when harvests are heaviest. Chef Andre commented that using the rooftop garden has caused him to rethink his menu and provide more seasonally appropriate cuisine rather than order out-of-

Andre Natera, head chef at the Fairmont Dallas, in a pumpkin patch in the rooftop garden

herbs for the kitchen. Chef Andre and his staff harvest fresh basil varieties, bay leaves, stevia, Mexican mint marigold, and other herbs as needed to use in their dishes. In addition to the freshness of the herbs, the garden supplies a much more extensive variety of herbs than may be available from suppliers.

The Fairmont Dallas also participates in the "Zip Code Honey" Program, run by the Texas Honey Bee Guild. Bee boxes are located on the rooftop, where the honey is harvested and then

Bee boxes in garlic chive boxes

season produce from many miles away. This practice is not only environmentally sound; it also provides a more flavorful, fresher taste to his preparations.

The rooftop garden has been in operation since 2007 and was partially created from a filled-in swimming pool. Savings in cost of produce are around ten thousand dollars per year. The restaurant still has to order some of the vegetables it serves; however, the garden provides most of the blended with honey from other nearby locations. A portion is returned to the restaurant for use in menu preparations.

Japanese Garden

While there are no absolute design elements in Japanese garden design, there are certain commonalities that make this design style distinctive. The Japanese garden style was originally borrowed from Chinese gardens and took on a distinctive flair with specific garden elements and styles. Now open for public enjoyment, Japanese gardens were traditionally reserved as private spaces for personal reflection for the royal families or as a place for meditation for Buddhist monks. Japanese gardens evoke feelings of tranquility and connection with the environment through the use of natural elements, creating mountains and valleys, rivers, streams, and beaches in microcosm to fit the scale of the site. The layout is asymmetrical in nature, requiring an artistic eye for scale and a vision for the beauty of the single element. For example, an ornamental tree or large shrub may be pruned into a twisted and gnarled shape to appear as a very old tree. It is very common to shape plants by specific pruning techniques to create unnatural forms from species that are adapted to heavy pruning. Similarly, rock forms are used to mimic mountain and plateau features found in nature. Beach and river forms may also be created with river rock and gravel. Seasons are accentuated with the use of evergreen and flowering species.

Apple tree blossoms (Photo by Andrew Edmonson)

Structure: Architectural elements, such as pagodas or teahouses, may be a part of the Japanese garden. Stone and architectural features are used to create the bones of the garden, with pathways connecting the various elements. There should be a sense of movement and discovery as one travels through the garden, creating moments of revelation. Height variances can be carved from the existing space to give the illusion of "mountains and valleys" that also aid in drainage and guide the visitor to discover the next garden element. Paths may be linear or curved but should have twists or turns and screens that reveal hidden vignettes and provide semi-hidden spaces.

Water: Water is an important element in Japanese gardens, even if it is an illusion. Dry streams can be created with gravel or river rock to give the illusion of flowing water, or actual ponds, streams, and waterfalls can be constructed. Bridges further enhance this illusion, creating crossings made from stone, wood, or large stepping-stones that connect created island and mainland land-masses. Water basins and decorative troughs are also used to punctuate Japanese garden spaces with water elements.

Suggested edible plantings: When selecting plants for a Japanese garden, it is important to consider form and texture, often using an interesting specimen that may be framed in a view or a single species that may form a mass ground cover or living wall. Prune trees to accentuate asymmetry or train to weeping forms to create interesting shapes. Plants for tea making are designated with an asterisk (*).

Artichoke
Asparagus
Bee balm (bergamot) (leaves)*
Blackberry (leaves)*
Chamomile (flowers)*
Fig
Garlic
Hibiscus (flowers)*
Italian stone pine
Kale
Kumquat
Leek
Lemon (fruit)*
Lemon balm (leaves)*
Lemon grass (leaves)*
Lime (fruit)*
Loquat
Mandarin orange
Mint (leaves)*

Mulberry
Pear
Pineapple sage (flowers and leaves)*
Plum
Pomegranate
Raspberry (leaves)*
Satsuma orange
Swiss chard

JAPANESE GARDEN PLAN: JAPANESE TEA GARDEN

An early twentieth-century split-level residence in central Austin is the site for this Japanese garden as a part of an overall renovation project. The house is situated on a lot with extreme grade changes, sloping from the front to the backyard. The landscape is designed to encourage movement and revelation throughout the space. Existing stairs and landings on the south side were reconstructed, and bamboo screens were installed along the walkways to screen an adjacent street and give a feeling of enclosure. The screens serve an additional purpose of "hiding" the teahouse in the lower yard, revealing the surprise as the garden visitor walks around the corner. The structure is nestled within existing shrubs that further act as a screen from the busy street.

An existing concrete platform in the northeast corner of the backyard is used as another garden feature, covered with a pergola and resurfaced with black gravel. A galvanized metal trough with a bamboo spigot is an added feature in this space, infusing an "East meets West" symbolism. Water features are used as periodic punctuation in various areas of the garden. There is a small bubbler fountain built into the bed in the small enclosed area on the upper level, and a water basin is nestled into the mint ground cover on the lower level, inviting the visitor to rest and delight in the sound of trickling water. Concrete lanterns containing candles are also spaced along the walkway to add natural nighttime lighting.

As many existing trees as possible were kept to screen the outside views and provide shade. Mint and elderberries are the most logical choices for areas that receive only filtered light. Many of the plant selections for this garden are commonly used for making tea. Fruit trees are also included in the garden and are designed to be pruned into asymmetrical "windblown" shapes. The plum tree is designed to be framed in the view from the teahouse for a beautiful view of plum blossoms in the spring. The persimmon provides a beautiful display of fall foliage for a lovely view from the elevated seating area, but fruit production will be limited because only one cultivar has been planted.

Pizza Garden

With the resurgence of outdoor kitchens, the outdoor pizza oven has become a popular way to share good food and fun times with friends and family. Gardeners can combine their love of gardening with a theme, such as pizza toppings, to create an even more personal experience. While raising the wheat for the crust and milking cows or goats to make mozzarella may be a little beyond the scope of most family gardeners, it may not be out of the realm of possibility for gardeners to brine their own olives, pickle artichoke hearts, and prepare homemade marinara sauce from the tomatoes and basil in their garden to top their pizzas.

Structure: A pizza garden can cover the entire landscape or only a portion of a more extensive garden and can be designed to suit any theme that works for the style of the garden. There are no rules of design style except that the plants should have a proximity to the oven to make the theme meaningful.

Suggested edible plants: The following plants can all provide ingredients for pizza toppings:

Artichoke
Asparagus
Basil
Eggplant
Garlic
Italian stone pine
Leeks
Marjoram
Olive
Onion
Oregano
Parsley
Pepper
Spinach
Tomato

'Navaho' Blackberry

6 - Elephant Garlic

'Fuyu' Persimmon

Concrete Lantern, Typ.

20 - Mint

Water Basin

3 - 'Aurea' Elderberry

'Methley' Plum

Fill in to Grade with
Black Tejas Gravel on
Existing Concrete Pad

Rough Cedar Arbor

Plant Water Lilies in
Galvanized Metal Trough with
Recirculating Pump

5 - 'Cow Horn' Okra

3 - Bee Balm

13 - Oregano

7 - Lemon Thyme

7 - 'UC-72' Asparagus

Decomposed Granite Walkway
with Chopped Block Limestone
Treads

Upright Rosemary with
Border of 5 English
Thyme

Tea
House

Residence

Existing Wall

Existing Concrete
Retaining Wall, Typ.

6 - Lemon Balm

6 - Compact Camellia
Sawn Limestone Pavers

'Key Lime' Basil Pot

6 - Upright Rosemary

8 - Lemon Grass

2 - Jerusalem Artichoke

4 - 'Fordhook' Swiss Chard -
Border with 20 Strawberry

Bamboo Privacy Screen
with Red Noodle
Asparagus Bean

3 - 'Purple Italian Globe'
Artichoke

5 - Chocolate Mint

Seed Chamomile around
Accent Boulder

1 - 'Hopi Red' Amaranth

Glazed Pot with Bubbler - Set
in Black Tejas Gravel

3 - 'Bright Lights' Swiss
Chard

Mortared Oklahoma
Select Flagstone on
Concrete Base

Scale: 1/8" = 1'-0"

Japanese tea garden design

65

Pizza oven (Photo by Lord Koxinga)

GARDEN PLAN: PIZZA GARDEN

What is more Italian than dining al fresco, eating homemade pizza with ingredients from the garden? This garden is designed around an outdoor kitchen with a built-in pizza oven and arbor-covered seating area. A cable trellis behind the kitchen forms a living screen with trellised vines of pizza toppings, such as zucchini and tomatoes. The seating area is on a level lower than the raised beds. Olive trees are planted in a taller bed that is built into grade to retain the slope. Decomposed granite walkways around the bed perimeters provide access to the seasonal vegetables from all sides to allow easy harvesting.

Permanent, evergreen plantings include a bay tree planted in the corner formed by the patio and the walk, olive trees as a backdrop to raised beds designed for seasonal vegetable plantings, and year-round ground covers of oregano and thyme. In the spring garlic scapes will display along the walkway, and artichoke and fennel will provide winter interest. Cheery pots of basil are set around the deck for summer interest. Basil pots may be brought in the house during freezing weather, but they tend to decline in winter months and should be replaced in the spring after all danger of frost has passed.

Walled Courtyard Garden

Walled courtyards have been an element of garden design throughout history. Often incorporating water features and artwork, these gardens have been spaces for reflection and respite since antiquity. Within the walled garden, there is protection. Offering a place for solitude or gatherings with friends and family, these spaces are an extension of the private living space of the residence. Aside from the restorative benefits of a walled garden, there are a number of functional advantages. Garden walls are used to protect plants from foraging animals and harsh winter winds and to provide vertical support, creating microclimates that are more comfortable for both humans and more tender plant species.

A well-designed courtyard garden should be an extension of the adjoining architecture, harmonizing the indoor and outdoor spaces. Depending on the desired effect, pathways can be curved or straight. Hidden gardens and reveals can be

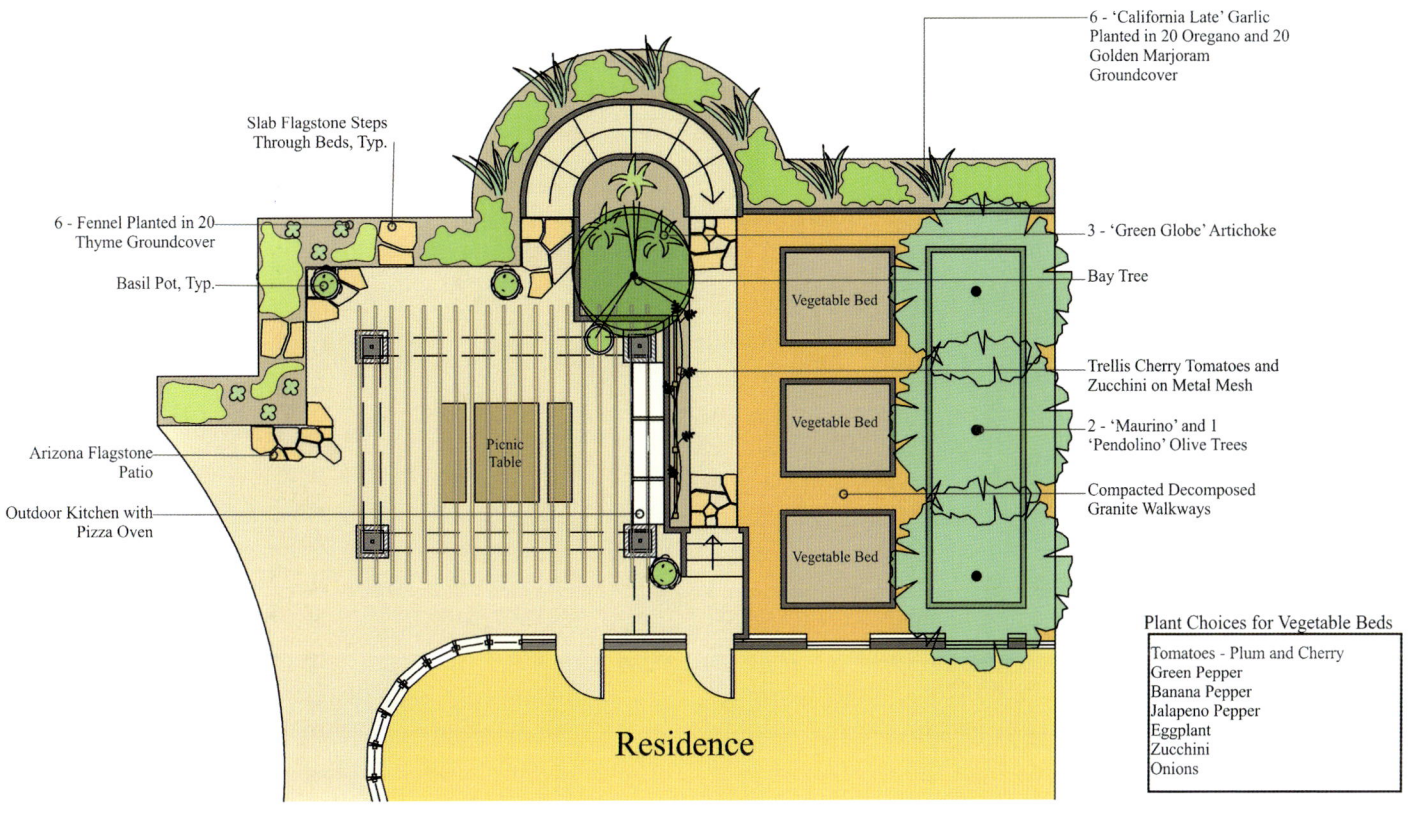

6 - 'California Late' Garlic Planted in 20 Oregano and 20 Golden Marjoram Groundcover

Slab Flagstone Steps Through Beds, Typ.

6 - Fennel Planted in 20 Thyme Groundcover

Basil Pot, Typ.

3 - 'Green Globe' Artichoke

Bay Tree

Vegetable Bed

Vegetable Bed

Trellis Cherry Tomatoes and Zucchini on Metal Mesh

2 - 'Maurino' and 1 'Pendolino' Olive Trees

Compacted Decomposed Granite Walkways

Arizona Flagstone Patio

Picnic Table

Outdoor Kitchen with Pizza Oven

Vegetable Bed

Residence

Plant Choices for Vegetable Beds

Tomatoes - Plum and Cherry
Green Pepper
Banana Pepper
Jalapeno Pepper
Eggplant
Zucchini
Onions

Scale: 1/8" = 1'-0"

Pizza garden design

Walled courtyard garden (Photo by Paul Shreeve)

designed to give the element of surprise and discovery, or paths may be aligned orthogonally. Paths laid out at right angles with parterre beds will create terminal and axial spaces that can be accentuated with fountains, artwork, and arbors. Windows and niches built into the walls add an architectural interest, and the illusion of adjoining outdoor rooms can be achieved with strategically placed mirrors that appear to be windows looking out to gardens beyond the wall.

Structure: Materials for garden wall construction are usually masonry, brick, stone, or stucco and may repeat the materials used in the construction of existing structures. Courtyards can also be created using screening hedges and wood or bam-boo fencing. If different materials from those in the existing architecture are used, they should be complementary in style and color. For example, a pink stucco wall might clash with a contemporary redbrick home. Columns, windows, and capstones can enhance the beauty and structure of the wall and tie the existing architecture to the outside space

Scale: The ratio between the height of the wall and the size of the space is an important consideration. A walled space can feel claustrophobic if the wall is too tall in relation to the size of the enclosed space, and a wall that is too short can seem inconsequential. Long, narrow garden spaces can be broken into smaller spaces with internal walls. While they should be compatible, internal walls do not need to be constructed of the same materials as outside walls. For example, "walls" can be made from trellised plants, giving the illusion of partitioned space.

Alignment: It is important to consider the existing architecture and try to enhance and repeat meaningful elements, such as entryways and repetitive architectural elements. Whether modern or English cottage style, the walled spaces should be an extension of the existing architecture. Axial elements should be aligned with existing doors and terminus views. Decorative fountains, trellises, and artwork can be used as a terminus for pathways, while central fountains and arbors can be used to define internal axes.

Plantings: Vertical growing space on a garden wall can increase the yield per square foot and provide an interesting element of a "living wall."

Hanging baskets and trellises can all be used to enhance garden walls while providing interest and vertical supports. Fruit trees can be trained as espaliers to create a more formal appearance. Although walled gardens offer an additional upright element, it is important to consider that walls also create shade, and north- and east-facing walls may cause shade pockets. While walls give protection from harsh winter winds, they also cast a shadow and shaded spots will take longer to warm.

Suggested edible plants: The following plants can be used in a walled courtyard:

Apple (espalier)
Bay
Chives (border plant)
Citrus (containerized or against south-facing wall)
Fig
Globe basil (border plant)
Grapevines
Loquat
Olive
Peach
Pear (espalier)
Persimmon
Plum
Pomegranate
Thyme (border plant)

WALLED COURTYARD GARDEN PLAN: BEESLEY RESIDENCE

This walled courtyard garden is designed as an extension of the living space, complementing the Mediterranean-style architecture with an adobe wall and heavy antique wooden gates. While there is an axial alignment between the entry gates and the front door, the design is informal, with an asymmetrical, curved flagstone walkway and seating area. Bed areas are located on the courtyard perimeter with a trellis and espaliered pear trees to soften the wall and add visual interest. An attached pergola over the kitchen windows and arbors over the front door and the main gate shade and accentuate the space and offer structures for vining plants, such as the large 'Cynthiana' grape trained on the pergola.

The courtyard walls provide some protection from harsh winter damage to citrus trees, which are placed in containers around the courtyard, and an attached greenhouse allows for easy access to move trees in and out when severe weather threatens. Citrus trees can be a concern in tight spaces because of their prominent thorns. A generous bed width from the walkways and a wide entry at the front door prevent well-pruned citrus thorns from snagging clothes and injuring unwary guests.

The plant selection and placement are specifically chosen for a purpose in this walled courtyard garden. Rosemary, lemon thyme, and citrus trees are selected for their evergreen and fragrant characteristics. Oregano and bay trees are selected for their evergreen characteristics as well, while

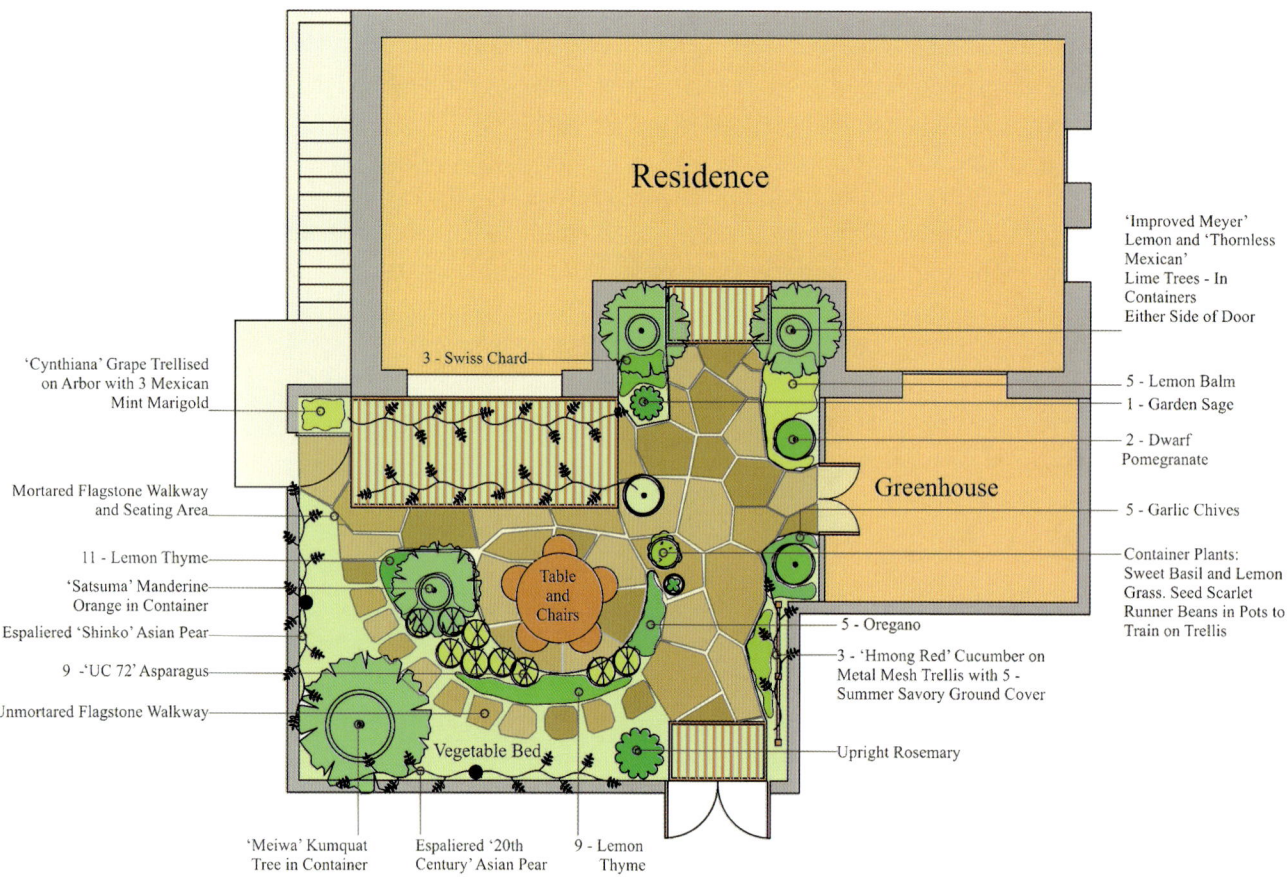

Residence

'Improved Meyer' Lemon and 'Thornless Mexican' Lime Trees - In Containers Either Side of Door

3 - Swiss Chard

'Cynthiana' Grape Trellised on Arbor with 3 Mexican Mint Marigold

5 - Lemon Balm
1 - Garden Sage
2 - Dwarf Pomegranate

Greenhouse

5 - Garlic Chives

Mortared Flagstone Walkway and Seating Area

Container Plants: Sweet Basil and Lemon Grass. Seed Scarlet Runner Beans in Pots to Train on Trellis

11 - Lemon Thyme
'Satsuma' Manderine Orange in Container
Espaliered 'Shinko' Asian Pear
9 - 'UC 72' Asparagus
Unmortared Flagstone Walkway

Table and Chairs

5 - Oregano

3 - 'Hmong Red' Cucumber on Metal Mesh Trellis with 5 - Summer Savory Ground Cover

Vegetable Bed

Upright Rosemary

'Meiwa' Kumquat Tree in Container

Espaliered '20th Century' Asian Pear

9 - Lemon Thyme

Scale: 1/8" = 1'-0"

Beesley residence courtyard garden design

they have the added benefit of being easily pruned to keep them in check and maintain a groomed appearance. Espaliered pear trees and topiary bay trees add elegance to the informality, a grapevine hangs gracefully from the pergola, and herb pots are placed strategically through the courtyard for added interest and warmth.

Low-Input Edible Landscaping

Edible landscape cultivation is inherently more demanding than more traditional ornamental plantings, both in time and materials. In general, edible landscapes require more water than xeric plantings, more nutrients and insect/disease controls than nonedible species, seasonal plant replacement, and a greater time commitment on the part of the gardener.

There are two ways to reduce labor and material resources in the edible landscape. The first option is to concentrate on edible plants that do not require as much care, and the second is to plant more compactly. Permaculture and square foot gardening are two techniques to accomplish the goals of an edible landscape within a low-input ecological system.

PERMACULTURE

Permaculture, a sustainable design model introduced by Bill Mollison and David Holmgren in the 1970s, is based on natural systems for both the built and the cultivated environment, with the goal of creating a sustainable human habitat.

Ecological patterns: Synergistic patterns between built elements, plants, and animals are designed to reduce both input and waste. For example, runoff from roofs is used to water plants; chickens are used to eat grubs and other insects; and plants provide fruit, shade, and windbreaks.

Design methods: When designing a permaculture system, the initial step is to take inventory of

Jujube fruit (Photo by James Beesley)

existing system input and output, for example, an analysis of water usage. Strengths and weaknesses are examined, and then a step-by-step modification plan is developed. Finally, the results of the modifications are analyzed and recorded to determine if further modifications are needed.

Water: There are three potential sources of water into a site: municipal water supplies, surface water from adjoining properties, and rain. Permaculture principles optimize the use of natural incoming water sources, creating swales and basins to catch storm runoff and recycled gray water and harvesting rainwater in cisterns. Gray water should be filtered through plants, sand, and gravel to purify it before use, and only the purest soaps should be used in the household. Municipal water use is minimized as much as possible in permaculture systems.

Suggested edible plants: Plants used in the permaculture model are selected for their ability to produce fruit without the need for an abundance of water, fertilizers, insect and disease control, or maintenance. Once established, this type of planting should require a minimal amount of resources, making it a more sustainable model. As a part of the permaculture model, plants are selected for their symbiotic relationships, especially in the use of legume ground covers to fix nitrogen for other non-nitrogen-fixing plants. Another example of potential symbiosis is using alliums and other pungent plants to deter insects, as well as flowers planted among vegetable plants to encourage pollination.

Low-input edible plant selections for Texas gardens: All of these plants will require consistent water until they are established and periodic watering during summer months and periods of drought. It is important to research each section of the bed to determine if the site is suitable for individual plants. Any of the plants in this book can be used in permaculture plantings if they are suited to the site. The following are selections that will require the least input.

Trees
Italian stone pine
Jujube
Loquat
Mulberry
Pear (some varieties more well suited, such as 'Kieffer')
Pecan
Persimmon
Pomegranate

Shrubs
Fig

Perennials
Blackberry
Dewberry

Herbs
Bee balm
Garlic chives
Lemon balm
Marjoram
Mexican mint marigold
Mint (in shade)
Oregano
Rosemary
Sage
Thyme

Vegetables
Amaranth
Garlic

Jerusalem artichoke

Okra

Southern peas

SQUARE FOOT GARDENING

Square foot gardening is an intensive gardening technique developed by Mel Bartholomew, in which plants are spaced within a 12- by 12-inch grid based on the desired ultimate plant spacing. For example, larger plants, like artichoke, will require a full square foot, and plants that only need a few inches of space, such as radishes, are planted up to sixteen to a square. As plants complete their life cycles, they are replaced, so the grids are almost always full. This technique is a particularly good choice for gardeners with limited space.

Structures: Frames 4 feet square are constructed with open bottoms, or weed barrier may be attached to the bottom. Usually weather-proof materials (1 by 6 inches) are used for the frame construction, and then 1-inch lattice pieces are attached to the frame to form the 1-foot-square grids. It is recommended that the boxes be placed in a way to allow 3-foot aisles in between. The 4-foot-square boxes allow easy access to the plants from outside the box without needing to step inside and trample the plants or compact the soil. Trellises attached to the boxes are used to accommodate space-hungry vining plants. Vining plants can take over a large area seemingly overnight, and keeping the vines from rambling over other plants in the bed will require vigilance.

Soils: Bartholomew recommends a mixture of three equal parts of compost, vermiculite, and peat moss for a lightweight soil mixture that holds water and nutrients. The only additional soil additive is a trowel of compost in each square when adding new plants.

Suggested edible plants: Plants are intercropped with a number of varieties. As one variety ends production, another is planted immediately to replace it, so the space produces optimally. It is important to plan carefully with square foot gardening to make sure the plants are rotated appropriately and family members are not planted in the same beds in consecutive seasons. Because the plants are closely spaced, it is also important to check plants regularly for diseases and insects.

PART III

Edible Plants for Texas

 # Trees

Tree Planting and Pruning

There are a few specific things to keep in mind when planting and pruning fruit and nut trees. The following notes and details will be helpful. Refer to the diagrams when planting or pruning the trees discussed in this section.

TREE PLANTING TIPS

1. Always call utility locators like DIGTESS to locate utility lines before digging.
2. If bare-root trees cannot be planted imme-diately, "heel in" the trees by laying them on their side and covering with clean, well-drained topsoil. Keep the roots moist but not wet.
3. Dig the planting hole twice as wide and the same depth as the root ball. The hole should have rough sides to give the roots nooks and crannies to cling to.
4. Loosen the roots if the tree is root-bound, place the tree with the graft union (appears as a bump a few inches above the root ball) a few inches above the ground, and plant the root ball at grade or just slightly above finished grade.

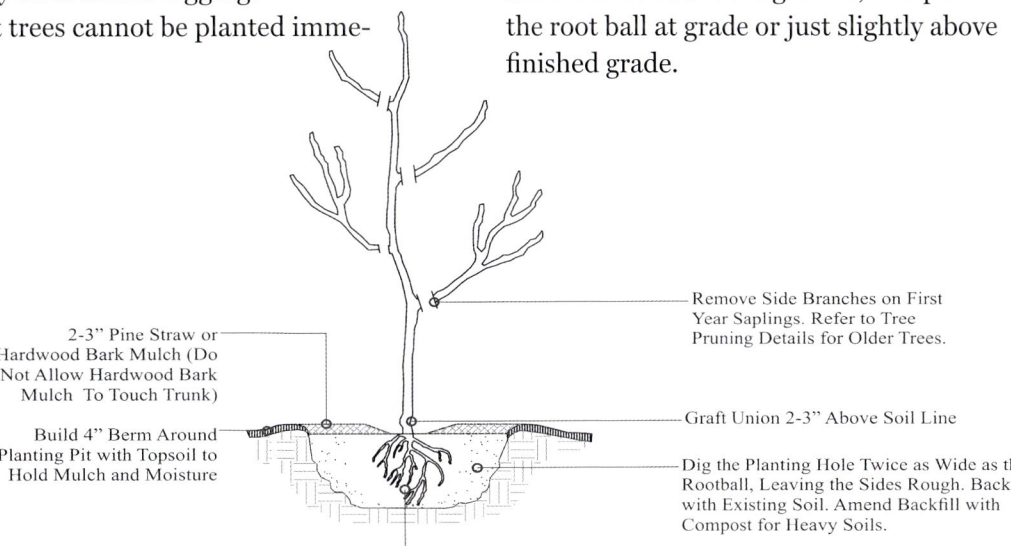

2-3" Pine Straw or Hardwood Bark Mulch (Do Not Allow Hardwood Bark Mulch To Touch Trunk)

Build 4" Berm Around Planting Pit with Topsoil to Hold Mulch and Moisture

Remove Side Branches on First Year Saplings. Refer to Tree Pruning Details for Older Trees.

Graft Union 2-3" Above Soil Line

Dig the Planting Hole Twice as Wide as the Rootball, Leaving the Sides Rough. Backfill with Existing Soil. Amend Backfill with Compost for Heavy Soils.

Guide for tree planting

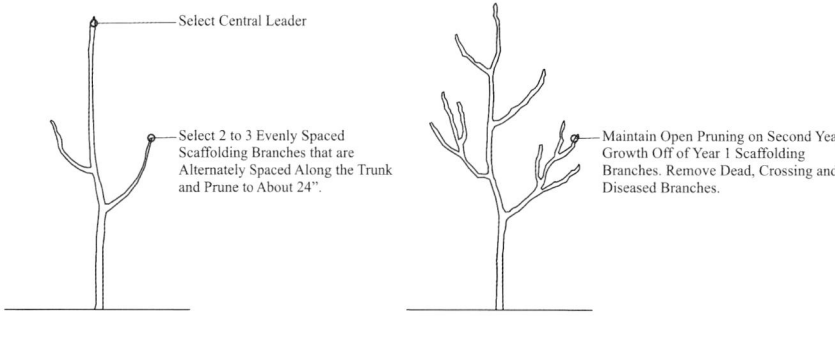

Select Central Leader

Select 2 to 3 Evenly Spaced Scaffolding Branches that are Alternately Spaced Along the Trunk and Prune to About 24".

Maintain Open Pruning on Second Year Growth Off of Year 1 Scaffolding Branches. Remove Dead, Crossing and Diseased Branches.

Year 1   Year 2

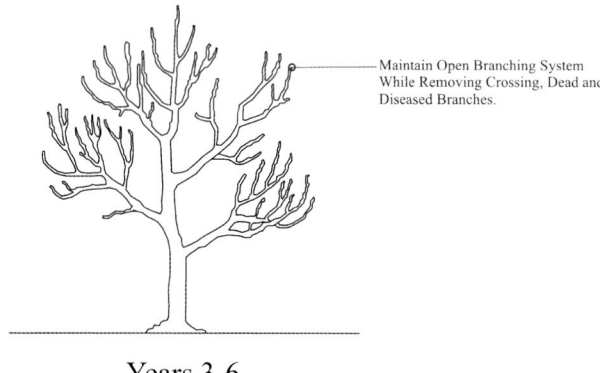

Maintain Open Branching System While Removing Crossing, Dead and Diseased Branches.

Years 3-6

Guide for central leader pruning

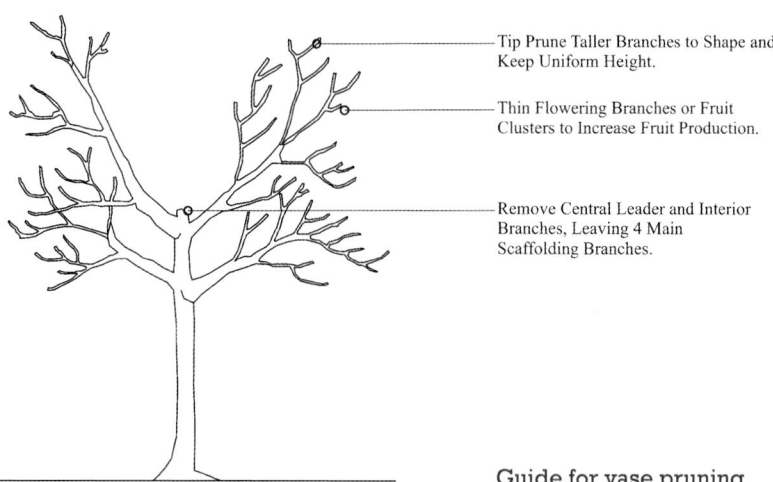

Tip Prune Taller Branches to Shape and Keep Uniform Height.

Thin Flowering Branches or Fruit Clusters to Increase Fruit Production.

Remove Central Leader and Interior Branches, Leaving 4 Main Scaffolding Branches.

Guide for vase pruning

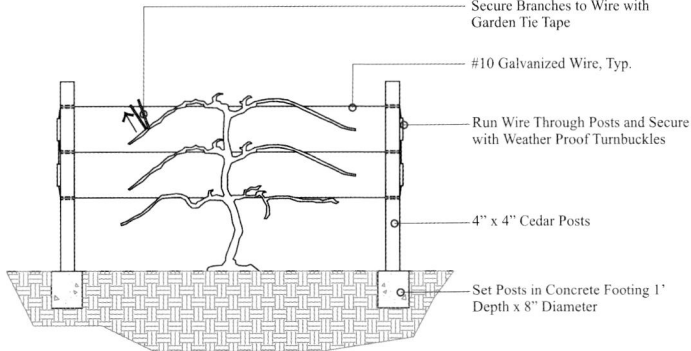

Secure Branches to Wire with
Garden Tie Tape

#10 Galvanized Wire, Typ.

Run Wire Through Posts and Secure
with Weather Proof Turnbuckles

4" x 4" Cedar Posts

Set Posts in Concrete Footing 1'
Depth x 8" Diameter

Lateral branch training

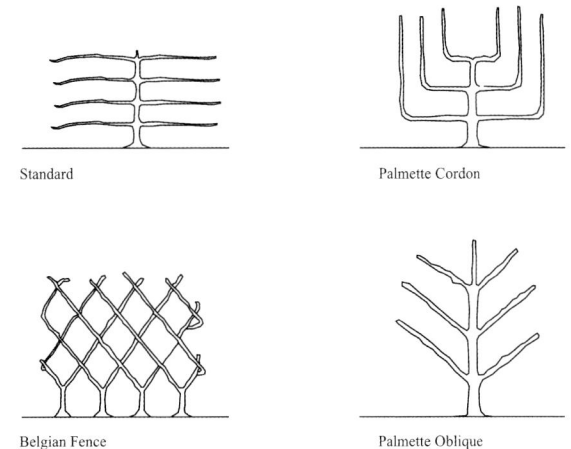

Standard

Palmette Cordon

Belgian Fence

Palmette Oblique

Guide for espalier training

5. Backfill the hole with the existing topsoil, watering in as you go to prevent air pockets. If soil is heavy clay, mix with compost before filling. Create a 4-inch berm around the planting hole to hold in water, and mulch the bare soil.

6. Trim back tree as needed per tree pruning details.

PRUNING TIPS

Pruning is an important part of regular tree maintenance. Following are a few simple rules to follow when pruning fruit and nut trees.

1. Do not prune when limbs are wet to prevent the introduction of fungal spores before the wound has healed over.

2. Prune branches at a 45° angle to allow water to run off.

3. Cut the branch flush with the branch or trunk it is growing from, leaving a small "collar" to allow the bark to grow over the wound.

4. Remove all dead or diseased wood, pruning diseased wood well below the apparent problem, as the disease may be present and not showing symptoms in lower parts of the tree. Clean pruning tools with rubbing alcohol between cuts and between trees.

5. Remove all sucker growth from the base of the tree.

6. Remove small and crossing branches from the interior of the canopy.

7. Maintain open air circulation. Do not be afraid to thin out branches.

8. Prune in the late winter when trees are dormant. (See specific information about olive tree pruning in that section.)

9. Prune individual flower and fruit clusters to encourage growth of remaining fruit.

Shade Trees

ITALIAN STONE PINE
Pinus pinea

✷ **DESIGN USES:** These large, evergreen pine trees can be quite magnificent. They can be used as shade trees; however, their upright branching and umbrella shape do not offer the abundance of shade that a pecan tree would. Italian stone pines also make good windbreaks for larger properties and orchards. Do not plant under power lines or where they will shade other plantings.

Italian stone pine (Photo by Andrew Edmonson)

Description

Italian stone pine or "umbrella" pine is an evergreen pine that has been in cultivation for over 6,000 years for its edible pine nuts. These large pine trees make a wonderful windbreak for larger properties or may be used as specimen trees on smaller lots. This very unusual pine is the only tree in the subsection *Pineae*. It has a unique form, essentially shaped like an umbrella, with a short trunk and a broad, rounded canopy. The needles differ on juvenile and mature trees. In the first 5 to 10 years the tree produces small, bluish-green, tufted needle clusters. The mature tree produces long, supple, bright green needles. The reddish-brown bark is thick and fissured. Italian stone pine produces ovoid-shaped cones that are green until they ripen, which takes 3 years, and then they turn brown. The pine nuts are ready to harvest at this time. Year-old seedlings are widely sold as small potted Christmas trees when they are small, but be sure to give this tree plenty of space. It can grow 40 to 80 feet in height with a spread of 40 to 60 feet.

Cultivation

Italian stone pine is originally from the Iberian areas of the Mediterranean and is well suited to many areas of Texas with loose, well-drained soils. Areas with hard, clay soils are not a good choice for this tree. Also, winter hardiness is questionable for regions north of zone 8. The Italian stone pine is drought tolerant and can be harmed by too much water; however, it should be watered on a regular basis for good nut production. Just do not allow the soil to stay wet around the roots. This tree does not require an abundance of fertilization. Amending the soil with compost along with foliar feedings of fish emulsion and seaweed is sufficient to keep Italian stone pines healthy.

OLIVE
Olea europaea

❋ **DESIGN USES:** This small, evergreen tree with silver or gray foliage can be planted to give a Mediterranean look to gardens in southern Texas counties.

Description

The olive tree has been grown for its oil and fruit for thousands of years in the Mediterranean region. Spanish missionaries brought the trees to Mexico and California, and the widely cultivated

Bella Vista olive orchard, Wimberley, Texas

'Mission' variety is named for their efforts. Olive trees are evergreen with lanceolate gray-green leaves. Young bark is green, turning gray as it matures. It usually takes about 5 years to produce fruit. The fruit is a drupe and is borne on the tree primarily in alternate years. Cream-colored male and female flowers are both produced on the tree; however, fruit production is increased if two or more varieties are planted. Olive trees are primarily pollinated by wind.

Olives have a very limited range in Texas. They are susceptible to freeze damage at temperatures below 17°F and will freeze to the ground when temperatures reach 12°F. Even though they are able to grow back from the roots, this can result in a weakened, shrubby-looking tree. If temperatures should fall below 10°F, the tree may die completely. Watering deeply will help olive trees survive freezing temperatures. According to Jim Henry of the Texas Olive Ranch, located southwest of San Antonio, Texas, winter rains encourage fruit production and help protect against hard freezes.

A limiting factor for olive fruit production is that olive trees require warm days and cool nights to undergo vernalization for fruit bud development, or they will not set fruit. The range for olive production is limited to a few of the more southern counties of Texas. Map 5 shows the range for olive production.

Olives cannot be eaten fresh because of a bitter glucoside and require additional effort to produce an edible fruit. They can be processed in a brine or lye solution to remove the bitter taste. There is no varietal difference between green and black olives; it is just a matter of how long they are left on the tree to ripen. The green olives are harvested sooner than the black ones.

Cultivation

Most olives are grown from softwood cuttings, so they mature on their own rootstock. Occasionally, cuttings are grafted to another type of rootstock. In this case, be careful not to cover the graft with soil, or the rootstock variety will sucker at the base of the trunk. Olive trees will tolerate a wide variety of soils as long as they are well drained. Heavy clay soils will not allow water to drain well enough for olive trees. The pH requirements vary from 5.5 to 8.5.

Planting: Plant olive trees in a sunny, well-drained site, following the diagram for tree planting.

Watering: In the summer months, olive trees will need deep watering from a bubbler or a soaker hose, making sure the water soaks deeply into the soil once or twice a week, depending on how hot and dry the weather is. Olive trees require more water and fertilization in the alternate years when they are producing more heavily, although they will commonly produce some olives each year. 'Arbequina' is a variety that tends to produce consistently every year, depending on the weather.

Fertilizing: Olive trees do not require fertilization beyond the basic organic soil preparation and foliar feeding. No fertilizer should be added after midsummer to encourage fruit production rather than vegetative growth.

Pruning: When the trees reach 4 years of age, it is time to begin the pruning process. Large interior branches will be removed over a few

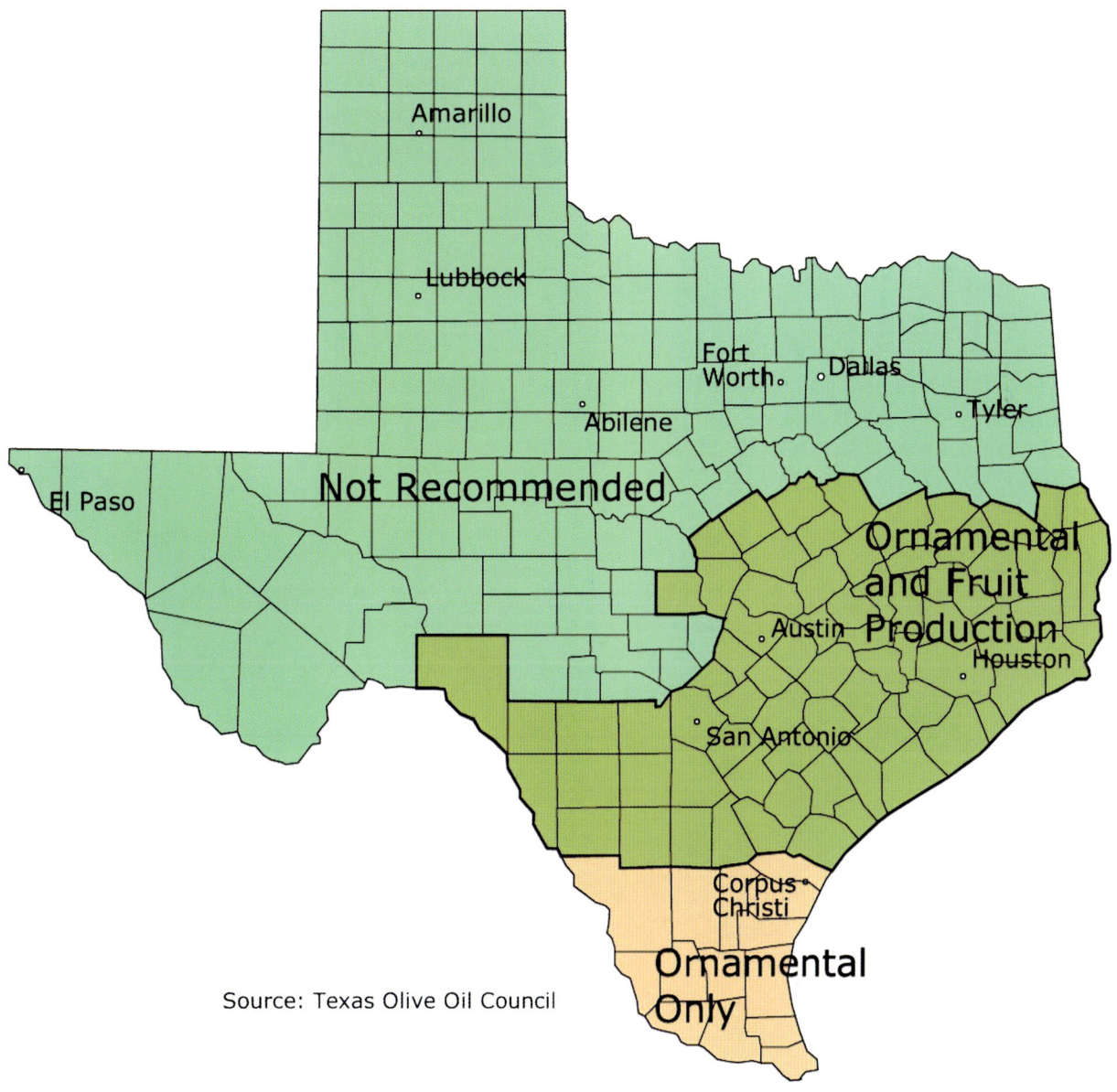

Amarillo

Lubbock

Fort
Worth

Dallas

Abilene

Tyler

El Paso

Not Recommended

Ornamental
and Fruit
Production

Austin

Houston

San Antonio

Corpus
Christi

Ornamental
Only

Source: Texas Olive Oil Council

Map 5: Texas counties suitable for olive fruit production

years' time, leaving three or four main scaffolding branches. This method is referred to as "open" or "vase" pruning, as shown in the diagram.

Because olive trees are evergreen, there is not a dormant season to prune, so the best time to prune is in the spring. Trees may be pruned more severely in alternate years when they are not producing as much fruit. Spring is also a good time to "thin" selected flowering branches to send more energy to the remaining branches. Tips may also be pruned at this time, keeping in mind that severe tip pruning of outer branches can produce a flush of new growth, sending energy into vegetative growth rather than fruit production. The overall goal is to prune out enough of the branches to allow light and air to reach as much of the remaining branches as possible. Olives must receive an abundance of sunlight to produce fruit. As recommended for any fruit-producing tree, remove sucker growth, dead wood, and cross-branching.

Protecting from freezes: To protect olive trees from freezing temperatures, use lightweight floating row covers, which can be purchased at most garden centers. Loosely wrap the trees with the cover, and tie it securely around the trunk. This will work for the first few years of the tree's life, until it becomes too large. At that point, it may be able to withstand the severe cold a little better. It is best to have these covers on hand before freezes are predicted, as temperatures can fall very dramatically in Texas. There is some question whether the degree of freezing temperature is as influential as the severity of temperature fluctuation in causing freeze damage in Texas olive trees.

Varieties

Arbequina A small tree with a weeping, compact form. Self-fertile and does not require a pollenizer but will perform better in the company of other olive trees. Fruit ripens from green to black in October or November.

Arbosana A good variety for smaller areas. Grows only 12 to 15 feet in height. Requires a pollenizer, and 'Arbequina' is a good choice. Begins producing in 3 to 5 years and hardy to 14°F.

Barouni A small tree, growing to around 15 to 20 feet. Developed in Tunisia. Fruit commonly grown as a table olive rather than for oil extraction and ripens to a reddish-purple in October or November. More resistant to freezing temperatures and extreme heat than some other varieties. A reliable producer, even in off years, especially with cross-pollination.

California Mission Can grow to 30 feet in height. Self-fertile. Vigorous and produces a good yield of deep purple fruits that ripen to black. Fruit ripens in October or November. Reported to be very hardy variety, surviving temperatures to 8°F, although that may not be accurate if temperatures drop dramatically. Fruit generally more bitter than other varieties, so grown commercially more for oil production than for the fruit.

Coratina A fast-growing, upright variety that produces large, elongated olives that ripen in November and December. Olives are ripe when the tips start to darken. Requires a pollenizer, and 'Frantoio' is a good choice.

Frantoio Can grow to 26 feet in height. Produces round, dark olives primarily used in oil production.

Maurino Medium sized with a pendulous habit. Produces small, ellipsoidal fruit that is black when fully ripened. Produces an abundance of pollen but is self-sterile and requires a pollenizer.

Pendolino A slow grower with a weeping characteristic. Ellipsoid fruit that is black when fully mature. Self-sterile but is a good pollenizer for other varieties. Crosses well with 'Maurino.

Olive Growing in Texas

Olives in Texas? Many have said that it could not be done because of the danger of freezes north of San Antonio and not enough chilling hours for fruit production south of San Antonio. Jack Dougherty is proving the naysayers wrong at the Bella Vista Olive Orchard in Wimberley, Texas. Jack and his family have a 27-acre "Italian farm," based on the traditional Italian farms described in the writings of Cato at the time of the Roman Empire. In addition to a vegetable garden and Red Angus cattle, they have an orchard of one thousand olive trees. They grow sixteen varieties, primarily the 'California Mission.' Jack recommends 'Arbequina' for the home grower, although this one may need more protection from snap freezes.

In Jack's experience it is not the cold temperatures in Central Texas that cause olive trees to die but the severe temperature fluctuations. Extreme cold snaps that can drop temperatures from the balmy 80s to below freezing in a matter of hours do not give the trees time to acclimate. The olives were first planted at Bella Vista in 1998. The orchard's first harvest was 3 years later in 2001. In 2002, 48 percent of the trees were lost to a snap freeze. Luckily, Jack and his family persevered and replanted. Today they have a thriving business, with an olive press where they cold-press their own olive oil.

PECAN
Carya illinoinensis

❀ **DESIGN USES:** Do not plant this large shade tree where it will overhang paved areas, structures, or walkways or under power lines. Take care not to plant where it will shade other plantings, as pecans get very large.

Byrd pecan orchard, San Saba, Texas

Description

The pecan is the State Tree of Texas, and many Texans have fond memories of sitting under these magnificent trees on a hot summer afternoon or gathering nuts on a cool autumn day. The stately branches of the pecan tree offer comfortable shade and an abundance of nuts (actually, they are a drupe fruit) for Texans. The trees can reach 80 to 100 feet in height, with trunk widths of 3 to 6 feet. The wood is very hard, and the trees can live for more than 200 years. Trees begin to bear in about 7 years and will bear heavier yields in alternate years. Pecans are monoecious, producing both staminate and pistillate catkins on the same tree. They are wind pollinated and cross-pollinate freely with neighboring trees. Cross-pollination is recommended for increased nut production. For cross-pollination, plant a type 1 (protogynous) variety with a type 2 (protandrous) variety to coordinate the correct blooming sequence. Pecans are well suited for many areas of Texas; however, there are some things to consider before planting:

1. Pecans have a large taproot that reaches deep into the ground in search of water. This limits their location to areas that have a deep layer of topsoil. Results will be disappointing if you attempt to plant a pecan tree on a rocky hilltop. The tree will grow until it hits bedrock and then will become stunted, and nut production will be minimal. Look around the neighborhood to see if there are any pecan trees. This will usually be a pretty good indicator of suitability.

2. Pecans are deciduous and messy and should be located where they will not overhang drives and walkways. They shed leaves, nuts, and catkins as well as sticky sap if they become infested with aphids. Care should also be taken not to plant too close to a structure or under utility lines.

3. There are many varieties of pecan trees, and choosing the right one can make all the difference in the success or failure of this tree in a landscape. Pecans are susceptible to fungal problems, especially in eastern counties. Pecans are also susceptible to aphid and pecan weevil infestations. It is best to choose varieties that have shown resistance to these insects.

Cultivation

Pecan trees are indigenous to North America. They are readily planted by squirrels and grow without much help from humans if they are located in a favorable spot; however, some cultural practices will increase the health of the tree and the resulting nut yield.

Planting: If possible, it is best to plant pecan trees in the fall or very early spring. This gives the tree some time to get established and adjust before the hot summers. Pecan trees are usually sold bare root. When purchasing container-grown trees, make sure the roots have not grown to a point where the taproot has been restricted and curled around the pot. It is best to order bare-root trees and have them delivered at the correct time for planting. If they are not going to go in the ground immediately, they will need to be kept in a moist soil media until they are ready to be planted in a permanent location. Generally, nonnative pecan trees are grafted onto native rootstock. The graft is easily recognized as a crook or a knot above the soil line, where the tree was planted at the nursery, which is usually distinctively different in color. This is the depth the tree should be planted in the ground. When planting the tree, care should be taken not to cover the graft with soil. If this happens, the native rootstock will sucker (send vegetative shoots from the roots).

Loosen the soil in the bottom of the hole to allow the taproot to grow. The tip of the taproot should rest at the bottom of the hole. The hole should be wide enough in diameter to allow feeder roots to spread, usually about 2 feet for bare-root or 5-gallon or smaller container-grown trees. Spread the lateral feeder roots growing from the sides of the taproot as you refill the hole. There is no need to add soil amendments, unless you are working with heavy clay, in which case you may need to work in compost with the clay. With other soil types, simply use the soil that has come out of the hole. Water the soil in as you fill to prevent air pockets from forming. Gently firm the soil in the hole with your foot while the hole is filled and soak thoroughly.

Watering: Trees will need to be watered deeply at least once a week and two to three times a week in the heat of the summer. This can be accomplished with an irrigation bubbler head or drip hose. Keep in mind that the water will need to soak all the way to the bottom of the root system when the tree is becoming established.

Fertilizing: Pecan trees can be fertilized organically with a mixture of well-decomposed compost and natural rock fertilizers applied at the base of the tree. Place the compost a few inches away from, not directly on, the trunk. The basic fertilization program described earlier should be used on pecan trees.

Pruning: As the tree matures, it will need to be pruned. All sucker growth, dead or diseased wood, and lower branches will need to be removed as the tree grows. Branches that crowd inside the canopy will need to be thinned to allow air circulation and light penetration.

Controlling insects and diseases: It is very important to keep tree debris cleaned up from below pecan trees to prevent insects and diseases from harboring in the fallen fruit and leaves. An important consideration in pecan tree pest control is choosing the right variety of tree. Fungal and aphid problems can be avoided by selecting resistant varieties; otherwise, repeated applications of fungicides to control scab and pesticides to control aphids and webworms will be necessary. Pecan weevils can be headed off by planting early varieties that can be harvested in September, breaking the weevil's life cycle. No pecan trees are resistant to the pecan weevil, and it is likely that this insect will become a problem if debris is not kept clean from around the tree. They can be easily transported from one area to another by human activity or mechanical means. Pecan weevils are a favorite food of bats, so bat houses are a good control measure. One of the benefits of an organic maintenance program is that beneficial insects are not harmed. A healthy, natural balance should help keep insect problems on your pecan trees to a minimum. Ladybeetles are very fond of aphids, a pest that is particularly fond of pecan trees that release a sticky sap that covers everything below the trees. Another biologically controlled insect is the webworm.

Varieties

Larry Don Womack with Womack Nursery in De Leon, Texas, recommends that only the most scab-resistant varieties be planted, especially if trying to grow the trees organically. The following are the varieties he recommends.

> **Caddo** Small nut (60 nuts per pound) with excellent kernel quality. Very disease resistant. Vigorous, upright growth. Protandrous.
> **Desirable** Large nut (39 nuts per pound) that cracks easily. Heavy producer. Disease resistant. Protandrous.
> **Forkert** Large nut (40 nuts per pound). Ripens in October. Protogynous.
> **Kanza** Small nut (72 nuts per pound). New USDA release. Very scab resistant. Matures in early September. Late pollen shedding, so good for northern counties as well as more southern ones. Shells easily. 'Pawnee' is a good cross-pollinator. Protogynous.
> **Oconee** Large nut (40 to 50 nuts per pound). Released in 1989 by the USDA and Texas, Louisiana, and Georgia. Early producer. Reported to have good scab resistance. Does not produce well if shaded. Protandrous.

Byrd Pecan Orchard

The Byrd Pecan Orchard, owned by John and Jimma Byrd, is located on the San Saba River near San Saba, Texas. The orchard has been in the Byrd family since 1944, and some of the trees date from that time. The Byrds have been growing pecans organically since 1995 and have been certified organic by the State of Texas since 2000.

According to John, there are a few things to consider before planting a pecan tree. First, are there any pecan trees growing in the area? Pecan trees have a long taproot that needs deep soil to grow. If the taproot hits bedrock, the tree will stop maturing properly. Second, consider the variety. Pecan trees cross-pollinate freely. Although both the male and female flowers are present on a single tree, they become available at different times, so it would be a good idea to plant more than one variety if you have the space. Avoid varieties that are more susceptible to scab and aphids. He also recommends early varieties to prevent damage from the pecan weevil, which generally invades in mid-August. Harvesting the nuts in mid-September might break the pecan weevil cycle. Any infested pecans should be destroyed, but breaking the life cycle will prevent further infestations the following year. John has had good luck with the following varieties:

> **Caddo** Small (67 nuts per pound), football-shaped nut with excellent kernel quality. Very disease resistant. Upright growth. Protandrous.

Kanza Medium-sized pecan (74 nuts per pound) with a thin shell for easy shelling. Very scab resistant and very early maturing. A new USDA release. Protogynous.

Lakota Heavy bearing and very resistant to scab with moderate resistance to aphids (59 nuts per pound). A new USDA release that is showing a lot of promise. Protogynous.

Mandan Matures in early September and is scab resistant (49 nuts per pound). Protandrous.

Nacono Very high nut quality, scab resistant, and upright habit (47 nuts per pound). Matures midseason. A new USDA release. Protogynous.

Pawnee Upright, pretty tree with a medium-sized nut (56 nuts per pound). Very early maturity. Has a mild susceptibility to scab. Resistant to aphids. Nuts have a good taste and are easy to shell. Protandrous.

The Byrds fertilize by applying a front-loader bucket of compost made from pecan leavings (the twigs, hulls, and leaves left from the cleaning process), mixed 50:50 with cow manure and some granite dust for trace minerals, around the drip line of the tree. They have inoculated their soil with both endo- and ectomycorrhizal fungi to aid the roots' nutrient intake. Zinc is applied if there are signs of a deficiency.

TEXAS WALNUT
Juglans microcarpa

Description
This variety of walnut is a Texas native. The nuts are small with a very thick, hard shell. The nuts are edible by determined wildlife but not worth the effort for humans. However, *J. microcarpa* does provide good rootstock for grafting more desirable walnut species. Some nurseries sell these grafted trees. 'Reda' and 'Fately' are two varieties readily available in Texas fruit and nut tree nurseries. Planting and cultivation requirements for walnut trees are the same as those for pecans.

Ornamental Trees

APPLE
Malus domestica

✽ **DESIGN USES:** Deciduous, ornamental trees with white blossoms and pink-tinged foliage in the spring produce pendulant clusters of yellow, green, or red fruit in the summer. Apple trees can be

Apple tree (Photo by Andrew Edmonson)

used in orchards, allées, and espaliers or as specimen trees. Dwarf varieties are useful as container plants or loose hedges. Columnar varieties are striking as geometric or containerized elements.

Description

The apple tree is a small, deciduous tree with beautiful pink buds that burst forth into white blossoms simultaneously with the leaves in the spring. It is a member of the rose family. Apples are not widely cultivated in Texas. They are generally considered to be a crop for the northeastern United States. Certain very flavorful varieties will do well with the proper care; however, they generally will not develop the red color for which apples are known. High temperatures suppress the production of red and yellow pigments. The most limiting factor for the would-be apple grower in Texas is the presence of cotton root rot in the soil. Alkaline soils are more likely to harbor this deadly pathogen. There are no resistant varieties, and there are no controls. So, it is important to have a soil test done before planting.

Cultivation

Well-drained, sandy loam soils are best for growing apple trees. Fine-textured, loamy clay soils will work if the drainage is good. To increase soil drainage, build up the existing soil with compost and other recommended soil amendments.

Order trees from a reputable nursery. It is also important to select varieties that have the proper number of required chilling hours for your area (see map 4). Urban areas are typically warmer and therefore have fewer chilling hours than rural areas. Check local conditions before selecting varieties. Fruit trees will usually be delivered at optimal planting time for your area and will usually be bare root. Inspect the tree roots, and make sure they have not dried out in shipping. If trees cannot be planted in their permanent site when they arrive, it is important to "heel them in" by laying the roots in hilled soil to keep them moist.

Planting: Site trees in a spot with full sun and good air circulation. Planting on an elevated site or hillside helps prevent frost damage, as cooler temperatures tend to linger in depressed areas. Full-sized trees should be planted at least 20 feet apart. Dwarf varieties can be planted closer and should be spaced according to mature height. As a rule, plants should be spaced at least two-thirds of their mature height from neighboring trees.

Dig the hole to the depth at which the trees were planted at the nursery, where the root crown flares from the trunk. Be careful not to cover the graft with soil. The graft will appear as a bump or curve above the root flare. Face the graft into the prevailing wind. Spread roots as the hole is filled so they lie on the soil at the depth at which they grew in the nursery, firming as you go. Water the tree in thoroughly to make sure there are no air pockets. Keep grass and weeds from below trees. Trees may be top-dressed with well-decomposed compost and mulched with 2 to 3 inches of finely shredded hardwood bark mulch or pine straw. Be careful that neither mulch nor compost touches the tree trunk.

Fertilizing: Top-dress apple trees with compost in the fall. If trees are fertilized in the spring with nitrogen-rich fertilizer, they will produce more new growth that is tender and more susceptible to fire blight. The tree will also put less energy into flower and fruit production.

Pruning: Apple trees need to be pruned in a modified central leader method for maximum fruit production. The primary objective is to open the tree to light and air. Trees are pruned to a central leader or trunk when they are first planted. In successive years, prune the trees during the dormant seasons to achieve an open structure built on this central leader. It is becoming more common to grow dwarf apple trees on a wire trellis system. Fruit should be thinned to one apple per cluster shortly after fruits begin to form. Clusters should be at least 3 to 4 inches apart. Trees produce fruit 4 to 6 years after planting. Dwarf trees tend to come into full production earlier than standard-sized trees.

Controlling insects and diseases: It is a good idea to spray apple trees with an organic fungicide in the spring as a preventive. Surround should be sprayed after petal drop to prevent infestations of woolly apple aphid, codling moth, and plum curculio. This product will also prevent sunscald on the fruit.

Rootstock: The selection of rootstock onto which the tree scion (a detached shoot or bud joined to the rootstock) is grafted will determine the size and disease resistance for the tree. Apple trees vary in height depending on their rootstock. The most dwarfing rootstock will produce trees that are about 6 feet in height, while standard-sized trees reach about 30 feet. Some nurseries allow specification of rootstock. Refer to the following list of rootstocks when selecting apple trees:

B9 Good fruit size. Needs staking or trellising. Resistant to root rots. Susceptible to woolly apple aphid. Grows 9 to 12 feet.

M7 Tree 50 percent of standard size. Fruit size is smaller. Virus-free rootstock adapted to heavy soils. High tolerance to fire blight and root diseases. Suckers badly.

M9 Tree 30 to 40 percent of standard size. Generally bears good-quality fruit. Roots are brittle, so tree needs staking or trellising. Susceptible to woolly apple aphid and fire blight. Moderate to heavy suckering. Some resistance to collar rot.

M26 Tree 40 percent of standard size. Good-quality fruit. Better anchored than M9, but some trellising or staking required in early years. Susceptible to collar rot. Extremely susceptible to fire blight.

M27 Most dwarf, growing to 6 feet, so makes a nice container tree. Fruit can be small. Must be staked or trellised. High susceptibility to fire blight.

MM106 Roots anchor well, so no trellising or staking needed. Fruit tends to be somewhat small, but productivity is high. Susceptible to collar rot and fire blight. Resistant to woolly apple aphid.

MM111 Tree 75 percent of standard size. Adapted to heavier soils. Most drought tolerant of the rootstocks. Does not require staking or trellising. Resistant to woolly apple aphid. Moderate resistance to fire blight and crown rot.

Varieties

Anna Light green to yellowish skin with a red tint. Fruit slightly tart and crisp. 300 chilling hours.

Arkansas Black Large, late-season variety with dark red skin. Good for cooking. Keeps well. 800 chilling hours.

Braeburn Late-season, crisp, tart green fruit with a red blush. Self-fertile, but better production if cross-pollinated. 700 chilling hours.

Dorsett Golden Golden delicious type with medium-sized fruit. Good companion with 'Anna' to increase pollination. 250 chilling hours.

Ein Shemer Prolific producer of large, yellow, crisp, tart fruit. Can be self-fertile. 350 to 450 chilling hours.

Empire Cross between 'Red Delicious' and 'McIntosh.' Sweet and juicy fruit. Good pollinator. 800 chilling hours.

Fuji Very flavorful, medium-sized, firm, reddish-orange fruit. Excellent pollinator. Self-fertile, but better production if cross-pollinated with another variety. 'Mollie's Delicious' a good choice for pollinator. 500 chilling hours.

Gala Medium-sized, golden-yellow fruit with a reddish-orange blush. Firm, juicy, sweet flesh. Stores well and is heat tolerant. 550 chilling hours.

Golden Delicious Medium-sized, sweet, yellow, crisp fruit. Good pollinator for 'Red Delicious.' 700 chilling hours.

Granny Smith Medium-sized to large, green fruit. Excellent for pies. Tart but can be eaten fresh. Self-fertile but 'Golden Delicious' a good pollinator. 400 chilling hours.

Holland Large, firm, sweet, juicy red fruit. Early variety. 500 chilling hours.

Jersey Mac 'Macintosh' type with medium-sized to large fruit with straw-colored striping. Medium sweetness with firm, crisp flesh. Ripens early July to early August. Resistant to cedar apple rust.

Jonagold Cross between 'Jonathan' and 'Golden Delicious.' Yellow with red blush. Great flavor. 700 to 800 chilling hours.

Mollie's Delicious Large, yellow fruit with a red blush. Flesh firm and juicy. 400 to 500 chilling hours.

Mutsu Large, crisp, yellow fruit. Very flavorful. Can be picked green for cooking or allowed to ripen to yellow for fresh eating. Pollinate with 'Gala,' 'Granny Smith,' or 'Fuji.' 600 chilling hours.

Northpole Columnar Large, 'McIntosh' type grows in a columnar shape. Good for container growing. Pollinate with 'Scarlet Sentinel.' 800 chilling hours.

Red Delicious (**Bisbee Spur**) Medium-sized, sweet, crisp fruit. Good pollinator. 700 chilling hours.

Apple in fruit, Natural Gardener, Austin, Texas (Photo by Andrew Edmonson)

Scarlet Sentinel Greenish-yellow fruit with red blush. Grows in a columnar shape. Good for container growing. Pollinate with 'Northpole Columnar.' 800 chilling hours.

White Winter Pearmain A very old variety with English heritage. Firm, yellow skin with fine-grained, rich, aromatic flesh. Good pollinator. 400 chilling hours.

Citrus Fruits

❋ **DESIGN USES:** Generally small, evergreen trees with intensely fragrant, white, waxy flowers that make lovely containerized plantings for most areas of Texas. Select dwarf varieties for container plantings. Citrus trees can be pruned into formal topiaries and are traditionally associated with formal gardens. The fragrant blossoms are a nice element when planted next to entryways or patios. Just be careful not to get too close to the thorns.

Description

Citrus trees are small, evergreen trees with very fragrant, white, waxy flowers. Their use is mostly limited to container growing in Texas because of their sensitivity to freezing temperatures. Citrus trees can be planted in the ground in zones 9 and 10 and with caution in zone 8b; however, they will need some protection if temperatures are expected to drop below freezing. The kumquats and 'Satsuma' orange may be exceptions to the container rule.

It is well worth the effort involved in protecting these beautiful trees to be able to grow the flavorful and abundant fruit they produce. Trees can be kept indoors in the winter if a sunny south-facing window is available, or they can be moved to a

Orange trees in terra-cotta pots lining walkway at the Palazzo Medici Riccardi, Florence, Italy (Photo by Giovanni Dall'Orto)

garage when freezing weather threatens. Keep in mind that this method involves a lot of heavy moving in and out of the garage as the weather changes. In light freezes, citrus trees can be wrapped with Christmas lights and covered with floating plant covers. Be sure to uncover when the freeze danger has passed or during warm daytime temperatures. Beyond winter protection, citrus trees are relatively easy to care for. Citrus trees are not afflicted with many insect or disease problems as long as they are healthy. Diseases and insects tend to attack stressed trees.

Cultivation

Citrus trees are not terribly picky about soil requirements. They just need lots of sun (8 to 10 hours a day) and good, well-drained soil. Citrus trees do not like wet feet and can be killed from being waterlogged.

Planting in the ground: Start by preparing a well-drained site (the south or southeast side of the house will be more protected from freezes). Improve heavy soils with well-decomposed compost. Mix the compost in thoroughly with existing soil to a depth that accommodates the root ball. Dig a hole to the depth of the tree's root ball. Gently spray off some of the growing medium from the root ball to loosen the roots, and set the tree in the hole, spreading the roots out while backfilling the hole. Firm the soil while filling to reduce the chance of air pockets. Make sure the tree is not sitting in a depressed area below grade. This could cause water to sit around the base of the trunk. Be careful not to cover the graft with soil. The graft will appear as a bump or curve above the root flare. Trees may be mulched with finely shredded hardwood bark mulch. The mulch should be at least 4 inches from the trunk of the tree. Any topdressing or mulch that comes in contact with the trunk of the tree could cause fungal problems. Eliminate all weeds and grass from beneath the tree.

Planting in containers: Fill containers with good-quality potting mix. Make sure both the soil and the container drain well. If the tree is small, it is sufficient to start out with a container that is just a little larger than the root ball. The size of the container will need to be "stepped up" as the tree grows; a mature citrus tree will eventually require a 20-gallon container. Placing the large containers on casters will make the trees easier to move to a protected area during freezes. Trees can be kept in smaller containers by pruning the roots when the plants become root-bound. Simply pull the tree out of the pot and trim off 2 to 3 inches all around the root ball. Replace the trimmed root ball into fresh potting soil.

Watering: Citrus trees need consistent watering but will not tolerate standing water around the roots. They will need to be watered more frequently if grown in containers, as the soil will dry out more quickly. Trees planted in the ground need to be watered twice a week for the first summer until they are established. Container-grown trees will need to be watered three to four times a week during the same period. As a rule, it is time to water when the top inch of the soil is dry. In both instances, cut watering back in the cooler months. Citrus trees can be killed quickly by too much water, so make sure the water drains away completely after it has soaked the roots. Trees that are stressed by too little water will drop leaves, and fruit production will be compromised.

Fertilizing: Citrus trees can be top-dressed with compost, keeping the compost away from the trunk. It is best to do this in the spring. Fall applications can cause new growth, which is more tender and susceptible to freezes. It is also a good practice to work in a yearly application of organic phosphorus. Feed trees with a foliar or liquid feeding of fish emulsion and seaweed on a monthly schedule. Greensand is a good source of micronutrients.

Pruning: Citrus trees do not require much pruning. A yearly removal of cluttered interior branches and a gentle shaping of the top is enough. Prune early in the spring to encourage new growth. As necessary for all trees, any dead or diseased wood and freeze damage should be removed as soon as it is noticed because it makes the tree more susceptible to damage from insects and disease. Sucker growth from the base of the trunk, usually growth from the rootstock, should also be removed when necessary. A number of rootstocks are used in the United States, but 'Sour Orange' is consistently used in Texas because of its cold tolerance and resistance to cotton root rot and *Phytophthora* (a genus of water molds).

GRAPEFRUIT
Citrus × paradisi

Description
A grapefruit tree will grow to 20 feet if planted in the ground. It can be kept smaller if planted in a container. When opting for container-grown grapefruit, plant trees grown on dwarfing rootstock. Grapefruit trees develop a rounding top with spreading branches. Texas is famous for its 'Ruby Red' grapefruit crop, which makes up 80 percent of Texas' fruit and nut exports annually. The tree was developed as a hybrid of a pomelo or

Grapefruit tree, Austin, Texas (Photo by Andrew Edmonson)

shaddock in the West Indies and was first described in the mid 1700s in Barbados. Grapefruit trees like hot days and warm nights. Much of Texas is ideal for these conditions. Unfortunately, they are also very susceptible to freeze damage. Mature trees can withstand temperatures in the mid-20s, but the fruit will freeze at 27°F if left on the trees.

Varieties

Henderson/Ray Similar to 'Ruby Red' except the peel is more attractive and the flesh is even redder.

Oro Blanco A hybrid between a grapefruit and acidless pomelo. Sweet, white-fleshed fruit with few seeds and larger than the true grapefruit. Foliage is glossy green. Matures in late winter to early spring and is replaced by huge, fragrant blossoms.

Rio Red A virtually seedless grapefruit with large, fragrant, white flowers and pink-red pulp. Fruit and juice are very sweet. The rind has a reddish tinge.

Ruby Red Very good-quality seedless fruit with red flesh. Peel is yellow with a pink tinge, and rind is thin.

KUMQUAT
Citrus spp.

Description
Kumquat trees originated in China and have been grown in Europe and North America since the mid-1800s. The kumquat tree is not a true citrus, although it has similar fruit and habit and will

Kumquat fruit (Photo by Aconcagua)

grow on citrus rootstock. The tree has a shrubby, compact habit and will reach 15 feet if planted in the ground. The leaves are deep green on angled branches that are usually thornless. Flowers are white and sweetly fragrant. Kumquats usually keep well because of their thick skin. They are good for fresh eating or for making marmalade. Kumquat trees are hardy to 15°F and can be planted in the ground in more southern regions of Texas if they are located in a protected spot. Space kumquats 8 to 12 feet apart as single trees or as close as 5 feet apart for hedge plantings. It is best to use dwarf kumquats if planting in a container.

Varieties
Kumquats are distinguished by species rather than as cultivars.

Hong Kong (*Citrus hindsii*) Nearly round, bright orange to scarlet fruit about ¾ inch in diameter. Not very fleshy. Tree is thorny.

Marumi (*C. japonica*) Round to slightly ovate fruit about 1¼ inches long. Peel is golden-yellow with scant, acidic fruit. Tree reaches about 9 feet in height and is more cold tolerant than other species. Leaves are small and branches slightly thorny.

Meiwa (*C. crassifolia*) A round kumquat about 1½ inches in size. The peel is yellow-orange and very thick. The pulp is sweet to sub-acid and usually seedless. Leaves are very thick and rigid and fold inward. Best kumquat for fresh eating.

Nagami (*C. margarita*) Large, oblong fruit about 1¾ inches long. Tree grows to 15 feet in height. Fruit ripens October to January.

Meyer lemon (Photo courtesy Bay Flora Nursery)

LEMON
Citrus × limon

Description

The lemon is an evergreen tree with sweet, intoxicatingly scented white flowers. Trees can reach 20 feet if planted in the ground and usually about 10 feet if planted in containers. There are two basic cultivars of lemon tree, 'Eureka' and 'Lisbon.' 'Eureka' lemon trees have an open, spreading form and grow vigorously. Their flowers are white, tinged with purple. They produce fruit on terminal growth, predominantly in the spring and summer. 'Lisbon' is a thorny, more cold-tolerant tree. It produces fruit inside the canopy year-round. 'Meyer' and 'Ponderosa' lemons are thought to be hybrid crosses and are very popular lemon trees for Texas. When the fruit is ready to harvest, it will turn yellow. Fruit can be kept on the tree after maturity for months; however, it should be removed if freezing weather threatens (and the trees brought inside).

Varieties

Eureka Spreading, open form that is not very thorny. New growth and flowers tinged with purple. Bears year-round, but most fruit produced in spring through summer. Clustered fruits borne on outside of canopy. Medium-thick rind with few seeds. More susceptible to insect infestations and freeze damage.

Improved Meyer A cross between a lemon and a mandarin orange. Hardy to 25°F, making it hardier than other lemons. Produces large, yellow-orange, juicy, year-round fruit. The original 'Meyer' lemon was almost wiped out by a virus in California in the 1960s. This improved strain is resistant to the virus and has become very popular for its thin, edible rind, high volume of juice, and lack of tartness. Not readily available in markets because it does not ship well. Smaller than other lemon trees and makes an excellent container tree.

Lisbon More tolerant of cold, heat, and wind than other varieties. Has more leaves and is thornier than 'Eureka.' Blooms and fruits all year. Seedless with medium-thick skin.

Ponderosa Very large (2 to 4 pounds) fruit that is very juicy with many seeds.

LIME
Citrus spp.

Description

Limes are small evergreen trees, producing white, fragrant flowers. The fruit generally falls from the tree when ripe, and the trees are generally more sensitive to cold than lemon trees.

Varieties

Kaffir Small fruit with a strong lime flavor and rough rind. Leaves are hourglass-shaped with a sweet citrus scent and widely used in Thai cooking. Makes a good container tree but is very thorny.

Key Lime (**Mexican Lime**) Moderate-sized, bushy tree with white, fragrant blossoms. Small, round fruit (1½ inches in diameter) with a thin rind and seeds, but very flavorful. Very cold sensitive.

Limequat Hybrid between 'Mexican' lime and kumquat. Slightly more cold tolerant than other limes.

Persian Lime Medium-sized to large, nearly thornless tree with a spreading form. Up to 20 feet if planted in the ground but smaller in containers. Large fruit (2 to 2½ inches in diameter) almost always seedless. Has similar cold tolerance to that of lemons. Thin rind turns yellowish when the fruit is ripe.

Thornless Mexican Lime Medium-sized tree makes a great container plant. Dense foliage produces a shrubby effect. Thornless limbs with fragrant, white flowers. Fruit and flowers produced year-round. Small, nearly round fruit (1½ inches in diameter) with a thin yellow-green rind. Somewhat less productive than 'Key' lime.

MANDARIN ORANGE
Citrus reticulata

Description

Mandarin orange trees are more erect growers than other orange trees and have more willowy branches. Limbs may need to be supported if they become heavy with fruit. They tend to be alternate bearing (bear every other year) and are more cold tolerant than other citrus. In the citrus family, only kumquats are hardier.

Mediterranean Oblate, medium-sized, yellow-orange fruit with a yellow-orange, thin rind. Usually has 20 to 25 seeds. Matures in November.

Miho A new variety showing promise for Texas. Texas A&M has found this cultivar to be of good quality. Matures early. Hardy to 13°F.

Ponkan Large mandarins with orange rind and flesh. Flesh is very flavorful and has a "melting" quality. Ripens mid- to late December.

Satsuma Grows 4 to 6 feet tall and makes a wonderful container plant. Fragrant flowers bloom in March and April. Sweet, low-acid, almost seedless fruit matures late October. Recommended for growing statewide as a container plant by Texas A&M AgriLife Extension Service. Hardy to 26°F.

Seto Developed by Texas A&M along with 'Miho.' Like 'Miho,' has good-quality fruit, matures early, and hardy to 13°F.

ORANGE
Citrus × sinensis

Description

The orange tree is a thorny, small to medium-sized (up to 20 feet when planted in the ground) evergreen tree with fragrant, white blossoms. Orange

Mandarin orange tree, Barton Springs Nursery, Austin, Texas

trees have been grown in containers for hundreds of years. One of the most well-known examples is from a garden constructed in the 1600s: the orangerie at the Château de Versailles in France. The trees at this palace have been grown in containers and overwintered inside and under cover. The average homeowner will not be able to duplicate the Gardens of Versailles; however, a less ambitious installation can be very rewarding. Specify dwarf or compact varieties for container planting.

Varieties

Blood Orange Small fruit with crimson pulp that is very sweet and flavorful. 'Moro,' the most brightly colored blood orange, is the most commonly planted variety and is thought to be a cross between the pomelo and the tangerine. 'Moro' ripens late winter to early spring, so it will need to ripen outside in the winter. Hardy to 27°F.

Everhard A compact navel orange tree. Consistent heavy producer of good-quality, seedless fruit. The "navel" not as pronounced as in some navel oranges. This and 'N-33' are the most widely planted oranges in Texas orchards.

Marrs A small orange tree with medium-sized to large, sweet fruit. Usually seedless unless pollinated by another variety of citrus. Fruit is easily bruised. Pulp is low in acid. Matures in early October and tends to be alternate bearing.

N-33 Navel Orange Medium-sized to large, very sweet orange that is easy to peel and

Orange tree, Saint Mary Cathedral, Austin, Texas

seedless. Matures in September, making it a good choice for outside fruit development in southern areas of Texas. This orange is a limb sport (mutation of a limb with fruit characteristics different from those of the rest of the tree) from a 'Marrs' orange in Edinburg, Texas. Hardier than most orange trees and cold tolerant to 24°F. Fruit susceptible to splitting in late summer. Navel oranges are from a single mutation from a monastery in Brazil. The mutation causes "twins," one of which is dormant. This is where the "navel" comes from. They are seedless and sterile, so propagation must come from grafted cuttings.

Parson Brown Medium-sized fruit, similar in quality to 'Marrs,' but has seeds. Matures in September.

Pineapple Medium-sized to large fruit, somewhat flattened on the ends with a moderately thick, smooth skin that has a good orange color. Usually has between 15 and 25 seeds. Fruit quality good. Generally matures end of November but fruit can hold on the tree until February. Tends to be alternate bearing.

Valencia Medium-sized orange that is virtually seedless and has very good juice quality. Ripens in January. Must ripen outside in the winter.

TANGERINE
Citrus tangerina

These small, sweet oranges are easy to peel and separate into segments, and the 'Clementine' variety is seedless, making this little fruit a healthy snack for adults and children.

Varieties

Clementine Deep reddish-orange fruit that will keep on the tree longer than most varieties. Fruit comes free from the rind easily.

Dancy Fruit darker in color and larger than 'Clementine' but does not hold on the tree as well. Has a later maturity date so is more susceptible to freeze damage.

Changsha A seedless tangerine that has relatively good cold hardiness.

JUJUBE
Ziziphus jujuba

❁ **DESIGN USES:** Jujubes are small, deciduous trees with airy foliage. Use in orchard plantings or in allees. They are not useful as specimen or interior garden plantings because of their thorns and rather unkempt appearance. Jujubes are excellent choices for permaculture plantings. Plant where they can be easily maintained and suckers removed.

Description

Jujubes love hot summers, are drought tolerant, and can grow in alkaline soils. It is a wonder they are not more widely cultivated in Texas. This graceful, small, ornamental tree with a twisted branching habit and naturally drooping branches can grow 20 to 25 feet in height. The tree's dainty appearance can hide its harsh natural defenses. The new wood has thorns, and jujube trees can also be very aggressive in the garden, sending sucker growth out into the beds to form thickets of plants if they are not kept under control. With proper pruning and a watchful eye on this tree's natural tendencies, jujubes can become very pretty garden specimens. They are self-pollinating, but

Jujube trees, Itz family farm, Fredericksburg, Texas

two or more varieties should be planted if there is enough room in the garden. This will increase production and provide an extended season since different cultivars ripen at different times.

There are hundreds of varieties of jujube, which have been categorized for two different purposes, fresh eating and drying. The trees that have been cultivated for drying are not as good for fresh eating. When dried, the fruits are tinted red and have a datelike quality. The varieties grown for fresh eating have a sweet, applelike flavor and can also be used in recipes in place of apples. The varieties listed are most commonly grown in the United States and easiest to find.

Cultivation

Once established, jujubes are virtually care-free. They do not have insect or disease problems and thrive in our harsh Texas summers. The trees are long-lived. Some trees in China are 1,000 years old, and stands of jujubes are occasionally found at abandoned farmsteads across the Southwest.

Planting: Jujube trees should be planted in full sun. Jujubes can tolerate poor soils, but they will perform better in a well-prepared bed with good drainage. A simple topdressing of compost is appreciated; however, jujube trees will grow and produce without added fertilization.

Pruning: Trees do not need much pruning in the canopy; only crossing or dead limbs need to be removed. The major pruning will be from the base of the tree where the sucker growth originates from the wild jujube rootstock.

Harvesting: Jujube trees begin to produce fruit in the first few years. The ripening fruit will turn from green to yellow and finally to a reddish color when ripe. The fruit grown for drying can be picked in the yellow stage, but the fruit for fresh eating should be left to ripen on the tree. When drying, be careful to stop the drying process when the fruit is spongy. When left to dry for too long, the fruit will become hard and inedible.

Varieties

Chico Named for the Chico Research Station in California where it was selected and bred for its high sugar content. Fruits 2 inches long and ripen in September. Good for fresh eating.

Contorta Tree has twisted branches that give it a sculptural quality, especially in winter when the branches are bare. Mahogany-colored fruit ripens in summer. Makes a wonderful specimen tree, especially if lit and silhouetted against a wall.

Honey Jar Small-fruited variety with very juicy fruit good for fresh eating. Ripens late August to early September.

Lang Pear-shaped fruit 1½ inches long turns a mahogany color when it ripens in September. Best for drying, but can be eaten fresh when fully ripe. Needs a pollenizer.

Li Rounded fruits 2 inches long turn a mahogany-purple color when ripe. Chartreuse flowers appear later in the season than those of most fruit trees so not usually harmed by late freezes. Good for fresh eating or can be dried.

Shanxi Li Large-fruited variety good for fresh eating or drying. Ripens late August to early September.

Sherwood Fruit can be eaten fresh or dried. Have a very close resemblance to dates when dried. Tree a little taller and less thorny than others. Fruit ripens in October.

Tigertooth Very sweet fruit 2½ inches long. A good choice in areas with higher humidity.

LOQUAT
Eriobotrya japonica

❋ **DESIGN USES:** These ornamental, evergreen trees produce panicles of white flower clusters followed by pendulant clusters of yellow-orange fruit. Their form is somewhat umbrella shaped. Use as an accent, an anchor planting on the corner of a structure, or a dense backdrop to shorter plants.

Description

The loquat is a small evergreen tree that grows between 12 and 25 feet in height. The rounded crown is formed by exotic, bold-textured foliage. The thick leaves are 5 to 12 inches long and 3 to 4 inches wide. They are dark green on top and have a fuzzy, rusty-brown texture underneath. New growth can be reddish. White loquat flowers are fragrant and bloom in late fall to early winter, when few other things are blooming. The yellow-orange fruit is round, oval, or pear shaped depending on the variety, and each contains three to five large brown seeds that come free from the flesh easily. Loquat fruit is not commonly grown commercially for anything but local markets, because the soft flesh does not transport easily and the fruit does not have a good taste if it is picked before it is ripe. Fruits should be allowed to ripen on the tree before harvesting. These sweet, acid to sub-acid fruits are a real treat, and most loquats are very prolific and productive when other fruit trees are just beginning blossom formation.

The loquat is subtropical and should not be planted north of zone 8b. Loquats can be grown as container trees in other parts of Texas, but they will require very large (20-gallon or greater) containers and protection during hard freezes. Well-established trees can survive in temperatures as low as 12°F, but flower buds will be killed at 19°F. Mature blossoms and seeds will be damaged at 26°F, causing the fruit to die. Many varieties mature early in January and February. Late-ripening varieties produce fruit as late as May and are better for less temperate areas of Texas. Loquats are usually not troubled by many insects or diseases.

Cultivation

Some loquat trees are self-fertile, and some require another variety for pollination. It is always best to plant more than one variety if there is

Loquat tree

room. Loquats are primarily pollinated by bees. They tend to be alternate bearing, producing a crop every other year. Thinning fruits during early formation can help production in the following year. Loquats can take some shade, but the best fruit production will be achieved with at least 6 hours of full sun per day.

Planting: Loquats are not too picky about soil type, but good drainage is essential.

Watering: Loquats are more drought tolerant than many fruit trees, but regular deep watering is required for good fruit production. They should never have sitting water around the roots, so make sure water percolates well through the soil.

Fertilizing: Loquats are not heavy feeders. A yearly topdressing of well-decomposed compost along with a regular organic feeding program should be all that is required.

Pruning: Loquats do not require much pruning. A simple shaping of the crown and removal of any dead or diseased branches are sufficient. Some pruning at the bottom of the trunk to remove lower branches will create a pretty, small tree form.

Varieties

The following is just a sampling of the more common varieties. Loquats can be grown on their own rootstock or on quince rootstock to produce dwarf trees that bear fruit in 2 to 3 years. Seedling stock does not bear for 8 to 10 years. The dwarf varieties are preferable for container growing.

Advance Naturally dwarf form. Medium-sized to large, deep yellow fruit borne in large, compact clusters. Flesh is sweet, whitish to translucent, and melting. Resistant to fire blight. Self-infertile but pollinates well with 'Gold Nugget.' Ripens midseason.

Big Jim Large fruits (1¼ to 1½ inches in diameter) with excellent flavor. Trees very productive. Fruit ripens March to April.

Champagne Medium-sized to large, yellow fruits that grow in loose clusters. Skin somewhat astringent. White flesh melting and very juicy with a slight champagne flavor. Flowers grayish. Self-fertile. Ripens late.

Early Red Medium-large, orange-red fruits with a tough, acidic skin. Fruit sweet and juicy with good flavor. Self-fertile. Ripens in April.

Gold Nugget Large, yellow-orange fruit with a tender skin and sweet orange flesh. Self fertile. Ripens late.

McBeth Large, yellow, flavorful, and juicy fruit with few seeds. Self-fertile. Ripens early.

Vista White Small to medium-sized, round, light yellow fruit with pure white flesh and very sweet flavor. Ripens end of April to early May.

MULBERRY
Morus spp.

✻ **DESIGN USES:** Mulberry is a small to medium-sized deciduous tree that can be pruned as an ornamental tree. The low branching makes a nice climbing tree for children's gardens. The flowers are not showy; however, the white, red, and black fruit can be very interesting. The fruit can stain, so do not plant near paved surfaces.

Description
Mulberry trees are small to medium sized (usually 30 to 45 feet, but some reports of 70 to 80 feet) and deciduous. The three species grown for fruit production are black (*Morus nigra*), white (*M. alba*), and red (*M. rubra*). The name applies to the color of the bud rather than the fruit. The inconspicuous flowers are borne in the leaf axils of the previous year's growth, and the berries ripen over a period of time rather than all at once.

Mulberry trees have long had a reputation as a "trash tree." This is a harsh term for a tree that produces such tasty fruit. The nonfruiting cultivar of this tree was widely planted in previous years and has not proved to be a very desirable tree. The fruiting varieties, however, produce drooping boughs of flavorful berry-type fruit and can be a very beautiful fruiting tree for Texas. (The fruit is actually a collective fruit, not a berry, but the flavor and form are very similar to those of blackberries.) The berries will stain badly, so the tree should not be located over any hard surfaces or paths.

Mulberry trees have large leaves that can be lobed or entire and vary between the types of tree. Frequently, both types of leaves appear on the same tree. The trees are grown commercially for silkworm production. Mulberry trees are often trained into a weeping form, in which the boughs drape around the trunk, but pruning in this unnatural manner can stress the trees. The branches are often low to the ground and are a favorite for

Ripening mulberry fruit (Photo by Andrew Edmonson)

climbing. It is a wonderful summer pastime to sit in the branches and feast on mulberries.

Black mulberry: This mulberry is native to western Asia and is the smallest of the three species, usually reaching only 30 feet. It has a bushy habit but can be pruned to a small tree form. The black mulberry is the least cold tolerant (0°F to 10°F), but it should be fine in most parts of Texas. High humidity can cause fungal problems on black mulberry trees. Full sun and good air circulation will help prevent fungus from developing. When well located, these trees are long-lived and can bear for hundreds of years. They are also the most flavorful of the three, producing large, juicy fruits that have a good balance of sweetness and tartness. The leaves are large and thicker than those on the white mulberry, with a downy pubescence on the underside.

White mulberry: This native of China is the largest and most cold tolerant of these three mulberry species and can be somewhat invasive. Leaves are thin, glossy, and light green. This tree leafs out early in the spring before the other two species. White mulberry trees can produce white, lavender, or black fruit. The white-berried trees add a unique interest to the garden.

Red mulberry: The red mulberry can grow to be very large (up to 70 feet in deep, fertile soil) but rarely lives more than 75 years. The leaves are similar to those of the white mulberry. Fruit is deep red to almost black and flavorful.

Cultivation

Mulberry trees produce best in full sun. They should be spaced according to height, at least 15 feet apart and where they will not overhang any hard surfaces or paths. Mulberry trees respond well to deep loam. They will survive in calcareous soils but will need organic amendments for the best fruit production. Soil should drain well. Mulberry trees will need regular irrigation for good fruit production. They do not require special feeding; a yearly topdressing of compost and basic organic fertilization will be sufficient. Prune to shape and remove dead or diseased wood. Mulberry trees tend to "bleed" sap when they are pruned. For this reason, it is best to prune live wood during the dormant period.

Varieties

Black Beauty Black mulberry tree that reaches about 15 feet in height. Sweet, juicy fruit that ripens late May to early June.

Collier A cross between white and red species

that makes a medium-sized, spreading, very productive tree that reaches a height of 20 to 40 feet. Medium-sized, purplish-black fruit with a good flavor.

Geraldi Dwarf Similar to 'Illinois Everbearing' and 'Collier' but grows to only 6 feet in height. Leaves and fruit the same size as the those of standard varieties, but form is more compact.

Illinois Everbearing Cross between red and white mulberries that produces black fruit 2½ inches long that is very good for fresh eating. Produces over a long time.

Pakistan Originated in Pakistan and produces fruits 2½ to 5 inches long. Breaks dormancy early and can be damaged in late-spring freezes. Large, glossy, heart-shaped leaves. Trees usually 20 to 30 feet in height but can reach up to 70 feet. Sweet, firm berries that tend to stain less than other mulberries. Ripens late July.

Persian Mulberry A black mulberry that grows 25 to 30 feet in height and has a dense canopy of heart-shaped leaves. Fruit about 1½ to 3 inches long.

Shangri-La A white mulberry that reaches about 20 feet in height and produces large, plump, black berries.

Tehama Very sweet, white fruit 2 to 3 inches long. Fast grower that can reach 30 to 35 feet in height. Breaks dormancy early and can be susceptible to late frosts.

PEAR
Pyrus communis

❋ **DESIGN USES:** Deciduous, small to mid-sized tree that produces white flowers in the spring and has yellow or reddish fall color. Pears are very versatile in their landscape usage. Standard-sized trees can be used as small shade trees, growing 35 to 40 feet in height. Standard and semi-dwarf varieties are beautiful planted as allees, and dwarf varieties can be used as loose hedges or planted in containers. Semi-dwarf pears are particularly well adapted for espalier. All sized trees can be used as orchard plantings.

Description

Pears are excellent fruit trees for Texas with few management problems. The showy white blossoms are one of the first flowers of the spring. The three types of pear trees are European, Oriental, and Asian. The European varieties, such as 'Bartlett,' 'Bosc,' and 'D'Anjou,' are well known for their good quality; however, they are not well suited for some areas of Texas because of their high susceptibility to fire blight. The Oriental varieties are much better choices. The Asian varieties, often called "apple-pears" because of their applelike texture, are newer to the scene and have shown promise.

Pears are not self-pollinating, so two pear trees are required for fruit production. Pollination is done by insects. Trees need to be located within 50 feet of each other for good fruit production. Standard-sized pear trees reach about 30 feet in height. Semi-dwarfing varieties are about 60 to

70 percent the size of standard-sized trees. Dwarf varieties are usually 8 to 10 feet tall. Height is determined by the rootstock.

Cultivation

Well-drained, sandy loam soils are best for growing pear trees. Fine-textured loamy clay soils will work if the drainage is good. To increase soil drainage, build up the existing soil with compost and soil-loosening amendments.

Order trees from a reputable nursery. It is also important to select varieties that have the proper number of required chilling hours for your area

Pear tree in fruit

(see map 4). Fruit trees are commonly delivered at optimal planting time for your area and will usually be bare root. Inspect the tree roots and make sure they have not dried out in shipping. If trees cannot be planted in their permanent site when they arrive, it is important to "heel them in" by laying the roots in hilled soil to keep them moist. It is best to select varieties with good fire blight resistance.

Planting: Pear trees prefer well-drained soil and a spot with full sun and good air circulation. The most severe threat faced by pear trees is fire blight. Plant resistant varieties, especially in more humid and rain-prone areas. Prior to planting, testing should be done for the presence of cotton root rot (*Phymatotrichopsis omnivora*) in the soil. Pear trees are susceptible to this disease, and there is no treatment available. Both cotton root rot and iron chlorosis can be a problem for pear trees, especially in alkaline soils.

Full-sized trees should be planted at least 20 feet apart. Dwarf varieties can be planted much closer and should be spaced according to mature height. As a rule, trees should be spaced at least two-thirds of their mature height from neighboring trees.

Dig the hole to the depth the trees were planted at the nursery, where the root crown flares from the trunk. Be careful not to cover the graft with soil. The graft will appear as a bump or curve above the root flare. Face the graft into the prevailing wind. Spread roots as the hole is filled, so they lie on the soil at the depth at which they grew in the nursery, firming as you go. Water the tree in thoroughly to make sure there are no air pockets. Keep grass and weeds from below trees, at least 3 feet from the trunk. Trees may be top-dressed with well-decomposed compost and mulched with 2 to 3 inches of finely shredded hardwood bark mulch or pine straw. Be careful that neither mulch nor compost touches the tree trunk.

Fertilizing: Top-dress pear trees with compost in the fall. If trees are fertilized in the spring with nitrogen-rich fertilizer, they will produce more new growth that is tender and more susceptible to fire blight. The tree will also put less energy into flower and fruit production. Foliar applications of fish emulsion and seaweed are beneficial during the growing season. Iron chlorosis is sometimes a problem in pear trees. Symptoms are seen as interveinal lightening in the leaves. This problem is caused by an iron deficiency and can be treated with iron chelates. Addition of organic material to the soil helps make iron more available to the tree's roots.

Pruning: Pear trees can be pruned in a multiple leader or a central leader method. The multiple leader method allows for the possibility of losing one of the leaders to fire blight. It also increases the amount of fruiting wood in the early years. Pear trees tend to produce vertical branching, so it is a good practice to weight the branches; however, it is not essential. The branches can be weighted with bricks or some other object tied to a weatherproof rope. Weighting should be done in the spring after petal fall when branches are most supple. A 60° angle is recommended between the limb and the weighted branch. This opens the canopy of the tree and produces a greater yield. Weighting can also produce a very interesting sculptural form. Be careful not to overstress branches, which could cause them to break in a heavy wind.

Rootstock: The selection of rootstock onto which the tree scion is grafted will determine the size and disease resistance for the tree. Pear trees in Texas are commonly grafted on to *Pyrus communis* rootstock. Some rootstocks have been cultivated with parent stock of 'Old Home' and 'Farmingdale,' called the OHxF series. These rootstocks have been selected for fire blight resistance. According to Jim Kamas of Texas A&M University, they have had very poor results with the OHxF rootstocks. They are included for reference; however, other rootstocks will be preferable.

OHxF 40, OHxF 513, and OHxF 87 Well-anchored, semi-dwarfing, two-thirds standard size. Resistance to fire blight, collar rot, woolly pear aphid, and pear decline.

OHxF 97 Standard size but better production than seedling stock. Good results seen with Asian pears.

OHxF 333 Well-anchored, semi-dwarfing, one-half to two-thirds standard size. Resistance to fire blight, collar rot, woolly pear aphid, and pear decline.

Pyro 2–33 Tree 85 to 90 percent of standard size. Early fruiting and anchorage. Still being tested at the time of this writing.

PyroDwarf Tree 40 to 50 percent of standard size. Begins fruiting in 2 to 4 years. Good fruit quality and rooting. Still being tested at the time of this writing.

Pyrus betulafolia Standard size. Good root-stock for Asian pears in early testing. Good anchorage. Chlorosis is a problem.

Pyrus calleryana A very common rootstock for pear trees in Texas, producing standard-sized fruit trees. Has shown some disease resistance and drought tolerance.

***Pyrus communis* "Bartlett Seedling"** Standard-sized trees from seedling rootstock. Resistant to fire blight and pear decline. Vigorous growth and good production.

Quince Commonly used for dwarf rootstock in Texas to produce trees that are 9 to 19 feet in height. Fruit produced in 4 to 5 years. Resistant to collar rot, woolly pear aphid, crown gall, and pear decline. Will need to be trellised or staked, as anchorage is a problem. Chlorosis is a problem.

Thinning: Fruit should be thinned to one pear per cluster shortly after fruits begin to form and spaced 6 inches apart along limbs. To remove fruit, hold the stem between the thumb and forefinger while pushing the fruit off with the other finger. Thinning allows the tree to send more energy and nutrients to the selected pears and to produce large, healthy fruit.

Harvesting: Standard-sized pear trees may take up to 8 years to bear fruit. Dwarf trees tend to come into full production earlier than standard-sized trees, usually around 4 to 5 years. Fruit should not be ripened on the tree. Pears ripen from the inside out and will have poor quality if left on the tree until they are soft. Remove fruit when there is a slight change of color from green to a yellow tinge and they become a little less firm to the touch. Pears should be left to ripen in a well-ventilated place for 1 to 2 weeks.

Varieties

Ayers Good-quality pear without grittiness. Blight resistant. 400 chilling hours.

Chojuro An Asian variety that produces greenish-brown, round pears with some russeting to the skin. Ripens mid-August and is very firm. Leaves have pretty fall color. 500 chilling hours.

Hosui Medium-sized to large Asian variety. Pollinate with '20th Century.' Ripens early August. 450 chilling hours.

Kieffer Durable variety found at many abandoned farmhouses throughout the South. Large fruit that is hard and gritty with coarse texture. May be best used for preserves, although can be eaten fresh. Late ripening (September–October). Keeps well. Good pollinator. 400 chilling hours.

Le Conte A medium-sized oriental hybrid that is a consistent producer, similar to 'Kieffer.' Medium-sized fruit. Ripens August and September. Good resistance to fire blight and hot, dry conditions. 200 chilling hours.

Magness A hybrid cross between 'Comice' and 'Seckel.' Medium-sized oval, greenish-yellow fruit with some russet coloring. Soft juicy flesh without much grit. Resistant to fire blight. Not a good pollinator. 400 chilling hours.

Moonglow Fruit is juicy with a smooth texture. Ripens late August. Vigorous growth with good fire blight resistance. 500 to 600 chilling hours.

Orient Large, round, sweet fruit with yellow skin and white flesh. Texture is smooth and firm. Resistant to fire blight. Good for cross-pollination. 400 chilling hours.

Shinko Medium-sized to large fruit with a rich, sweet flavor. Ripens mid-August to mid-September. Most resistant to fire blight of all Asian pears. Stores well. 450 hours.

20th Century Asian Pear (**Nijisiki**) Called "apple-pear" because of similarity to apple texture and shape. Trees produce at an earlier age. Ripens mid-August. 400 chilling hours.

Warren Varies in size, but usually a medium-sized, long-necked fruit. Very sweet, juicy flesh with little grit. Ripens early August. 600 chilling hours.

PERSIMMON
Diospyros spp.

❀ **DESIGN USES:** Small, multi- or single-trunked deciduous trees grow to around 25 feet tall and wide. The fruits can persist on the tree after the yellow, orange, and red fall foliage has dropped, leaving the pretty orange to red fruits decorating the gracefully drooping branches. Makes a lovely specimen tree.

Description
The persimmon is a wonderful choice for Texas and is relatively care-free. There are two types of persimmon cultivars: astringent and non-astringent. The astringent fruits are almost inedible until they have ripened completely; otherwise, they will cause the mouth to pucker. The ripened fruits of both types are sweet and slightly tangy. The rind is firm but edible.

Persimmon flowers are not showy. Trees can be male or female, and some trees are both. Male plants can occasionally produce misshapen fruit from perfect (bisexual) flowers. Many persimmon cultivars set seedless fruit without being pollinated. If these trees are pollinated, they can produce fruit with seeds that are different in taste from those produced from nonpollinated flowers.

Cultivation
These trees are generally cold hardy for Texas; however, because of low chilling-hour requirements (100 hours), they may break dormancy early in the spring and the flowers can be damaged by later spring frosts. The leaves are killed at 26°F, but the trees are hardy to 0°F if they are completely dormant. If not completely dormant, they can be harmed by subfreezing temperatures.

Planting: Persimmon trees do best in full sun, planted where their taproots can reach deeply into the ground. They do particularly well in loam soils, but they can tolerate more alkaline soils as long as they are well drained. Trees should be spaced 15 to 20 feet apart. If persimmon trees are given plenty of sun and good air circulation, they should not have any problems with fungal diseases.

Watering: Persimmon trees do not need an overabundance of water, but they do require regular irrigation for fruit production. Trees should be watered deeply at least once a week and twice a week in hot summer months. Water should soak deeply in the ground and drain completely.

Fertilizing: Persimmons do not need much additional feeding. Apply a topdressing of well-decomposed compost in the winter (not in contact with the trunks of the trees). Excess nitrogen from the compost could cause fruit drop if it is applied in the spring.

Pruning: Persimmon trees can be pruned to form graceful small trees. Remove lower branches and "lift" the canopy as the tree grows. Remove suckering growth from the root flare and any dead

Persimmon tree, Natural Gardener, Austin, Texas

or diseased limbs. Persimmon fruit is borne on the outside of the canopy, and branches can become heavy with fruit and break in a high wind. Prune branches to develop a strong framework of main branches to hold the fruit-bearing branches. Maintain a vase shape, allowing outer branches to arch. Trees should be cut back to about 3 feet at the time of planting.

Harvesting: Fruits should be clipped from the tree with handheld pruners, cutting the fruit right above the calyx (the outermost parts of the flower). Astringent varieties can be ripened at room temperature, or they will keep for up to a month in the refrigerator. Non-astringent varieties ripen faster and do not store well. Both types should be handled carefully after they are harvested, as the fruits bruise easily. Fruits can be eaten fresh when they are ripe, or they can be dried or frozen for future use. When astringent fruits are dried or frozen, they lose their astringency. Dried persimmon fruit has a datelike consistency.

Insect and disease problems: Persimmon trees are not susceptible to many disease or insect problems. They can occasionally be infested with aphids and scale, which are symbiotic with ant infestations. Ants can be controlled effectively with certain species of beneficial nematodes if they become a serious problem. Many of our furry and feathered friends (deer, squirrels, opossums, and birds) enjoy persimmon fruit. These scavengers can be outsmarted by harvesting the fruit before it is completely ripe, or trees can be covered with fruit tree netting.

Varieties

There are more than 2,000 cultivars of persimmon, and experimentation with varieties could be very rewarding if there is room in the garden. The following are the more common varieties.

American A native persimmon that can grow very large and has a rounded form. Tree has dark, checkered bark that provides winter interest. Sweet, yellow fruit has jellylike texture. Ripens in September and November. Very cold tolerant. Flowers are male or female, so this variety requires a pollenizer. Astringent.

Chocolate Medium-sized orange fruit with a brownish, "chocolate"-colored tinge to the flesh. Fruit sweet and elongated. Ripens late October to early November. Good pollenizer. Astringent when seedless.

Eureka Medium-sized to large, bright orange-red fruit. A small drought-tolerant tree and more cold hardy than other varieties. Produces consistently, and fruit is good quality. Astringent.

Fuyu Medium-sized to large, deep orange, round, somewhat flattened fruit. The most commonly cultivated Japanese persimmon for fresh eating. Ripens in November. Very sweet, crisp fruits keep well and are non-astringent.

Great Wall Medium-sized, very sweet, flattened, orange fruit. Ripens in September to November. Tree grows to 15 feet in height and is very productive and cold hardy. Good fruit quality. Self-fertile. Astringent.

Gwang Yang Very large, seedless fruit can be eaten hard. Dwarf tree grows to 10 feet. Non-astringent.

Hachiya Large, deep orange fruit up to 4 inches with dark yellow flesh. Tree grows 12 to 15 feet in height. Fruit grown for market in the United States. Flavor is sweet and a good selection for drying. Ripens late October. Astringent.

Honan Red Small, thin-skinned, orange-red fruit with sweet flavor. Good for drying. Astringent.

Izu Medium-sized, burnt-orange fruit produced on semi-dwarf trees. Soft flesh with good flavor and texture but can revert to astringency. Ripens late September to mid-October. Female flowers only. Needs a pollenizer. Good producer. Primarily non-astringent.

Jiro Large, orange-red fruit similar in shape and size to 'Fuyu' fruit. Has good flavor and keeps well. Ripens late October to early November. Non-astringent.

Kyungsun Ban-Si Korean variety with large, showy, dark green leaves and rounded form. Deep orange fruits 3 inches across and soft when ripe. Good flavor. Astringent.

Miss Kim Dwarf variety a good producer with good cold tolerance. Ripens early October into winter. Good fall color. Astringent.

Rosseyanka Cross between American and Asian types that grows 15 to 20 feet tall. The leaves are similar to those of the American parent, and the fruit similar to that of the Asian parent. Orange fall leaf color. Astringent.

Ruby American Native Very large fruit ripens August to November. Pretty, ornamental form. Self-fertile. Astringent.

Saijo Small, elongated, yellow-orange, sweet fruit of good quality. Self-fertile, medium-sized trees bear consistently and hardy to 10°F. Astringent.

Tamopan Large, reddish-orange fruit with light orange flesh. Sweet when ripe. Ripens in November. Astringent.

Tane-nashi Large, cone-shaped fruit that is dark yellow to orange when ripe. Flesh yellow and sweet. Large, vigorous tree with a rounded form. Ripens late September. Astringent.

POMEGRANATE
Punica granatum

❊ **DESIGN USES:** Pomegranates can be grown as a large shrub or as a small multitrunked tree, depending on how they are pruned. The beautiful orange flowers on graceful arching branches make a dramatic display in either case. The pendulant orange-red fruit can persist after the bright yellow leaves fall in the autumn. Dwarf varieties produce full-sized fruit and make very attractive container plants that can be brought inside if needed to protect from freezing weather. Full-sized varieties can be used as a screening hedge, although they are deciduous.

Description
Pomegranate trees originated in the eastern Mediterranean regions and have been cultivated since ancient times. Spanish missionaries introduced the trees to the Americas in the sixteenth century. These trees prefer cool winters and hot summers and are very well suited to the more southern counties in Texas. Pomegranates can die to the ground or be killed at temperatures between 10°F and 18°F, so they may need protection in severe freezes. The hard-seeded and standard-sized varieties may be more cold hardy.

Pomegranate trees can be left as large, bushy shrubs or pruned to a small tree form by removing the lower limbs and frequent sucker growth. They generally grow between 15 and 20 feet in height, except the dwarf varieties that range from 3 to 7 feet. Aside from the edible fruit, pomegranates offer bright yellow fall color and showy, single or double, red-orange flowers early in the summer. The leathery orange to red fruit skin contains many crunchy seeds that are edible. These seeds are surrounded by a sweet, pulpy, red juice from which grenadine syrup is produced.

Cultivation
Pomegranate trees like full sun and well-drained soil. Other than that, they are not very fussy. Remove trees from the container and loosen the soil around the roots. Water trees until they are established every two to three days or whenever the top 2 inches of soil is dry. Make sure the water soaks deeply into the ground and drains away easily. Mulch trees to maintain soil moisture and protect from freezing. Pomegranates will appreciate a yearly top-dress with well-decomposed compost.

Pomegranate pruned into tree form, Austin, Texas

Controlling pests and diseases: Pomegranates are not usually bothered by insects and diseases, except occasional fungal problems. Planting trees in full sun with good air circulation should prevent this from occurring. If a problem develops, spray the leaves with a baking soda solution or other organic fungicide.

Varieties

Angel Red A relatively new variety with soft seeds and very sweet juice, contained in a bright red skin. Matures early September.

Granada Similar to 'Wonderful,' but flowers are bright red and fruit ripens in August, a month earlier.

Nana Dwarf Dwarf variety that grows 3 to 6 feet in height. Dark red fruits that are somewhat tart and smaller than other varieties. Semi-evergreen in mild winters.

Phil's Sweet Very sweet, medium-sized to large pink fruits with soft, less conspicuous seeds.

Russian Similar to 'Wonderful' but much more cold tolerant. Reported to withstand temperatures as low as -6°F.

Sweet A large pomegranate fruit with a light pink flesh. Ripens in September.

Utah Sweet Tree has a rounded form and grows 20 to 30 feet in height. Both self- and cross-pollinated. Sweet fruit bears seeds that are softer than those of many varieties. Flesh and skin are pinkish.

White Fruit with white pulp and red seeds. Striking when sliced open cross-wise. Very sweet and juicy. Ripens early September.

Wonderful A widely planted variety that grows to 18 feet. Sweet, juicy fruit has red skin and flesh.

Stone Fruits

✳ **DESIGN USES:** Deciduous, ornamental trees with showy pink spring flowers and summer fruit. Standard and semi-dwarf trees are very pretty as orchard plantings or lining walks or drives as allees. Dwarf varieties can be grown as a deciduous, loosely spaced hedge or in containers.

Description

The stone fruits (*Prunus* genus) include apricots, nectarines, peaches, and plums. Although trees in this family are known to be short-lived, they are worth the effort for their beautiful spring color and delicious fruit. In the spring, these fruit trees are covered with sprays of fragrant pink blossoms and are very striking when planted in allees. When used as informal plantings, they make beautiful backdrops for gardens. *Prunus* fruit trees require special attention to produce healthy fruit. Stone fruits are generally susceptible to similar disease and insect problems. To achieve good results, it is very important to check the trees regularly for insect pests and disease and to treat for problems preemptively if possible. Make sure all fallen fruit and debris are removed from the base of the tree to prevent insects and diseases from harboring in plant litter. Feathered friends enjoy the fruit as well, so use of bird netting can help ensure a good harvest.

Cultivation

Good soil drainage is very important when growing stone fruit trees. If the soil is too heavy, the water will not drain, and the trees will often develop fungal problems. This problem can be corrected by building up the existing soil with compost and other soil amendments.

Order trees from a reputable nursery. It may be necessary to check if they are grown on rootstock resistant to root-knot nematodes, a common problem in sandy soils. 'Lovell' is the suggested rootstock for soils that are not sandy. 'Nemaguard' rootstock is suggested for more sandy, acidic soils. 'Halford' rootstock has similar adaptability as 'Lovell.' Like 'Lovell,' it has no nematode resistance; however, this rootstock is more cold hardy and will adapt to heavy or high-pH soils. It is also important to select varieties that have the proper number of required chilling hours for your area (see map 4). Fruit trees will usually be delivered at optimal planting time for your area, generally in January or February, and will most likely be bare root. Inspect the tree roots and make sure they have not dried out in shipping. If the trees cannot be planted in their permanent site when they arrive, it is important to "heel them in" by laying the roots in hilled soil to keep them moist.

Planting: Site the trees in a spot with full sun and good air circulation. Full-sized trees should be planted at least 20 feet apart, and dwarf varieties, 12 to 15 feet apart. Dig the hole to the depth the trees were planted at the nursery, where the root crown flares from the trunk. Be careful not to cover the graft with soil. The graft will appear as a bump or curve above the root flare. Face the graft into the prevailing wind. Replace the soil around the roots of the tree and firm the soil in the hole. Water the tree in thoroughly to make sure there are no air pockets. If trees are planted in lawn areas, ring the trees to keep grass out of the drip line of the tree, as turfgrasses will compete with the trees for nutrients and water. Mulched areas should be as large as the tree canopy if possible. Place 2 to 3 inches of mulch around each tree, making sure the mulch is not in contact with the trunk.

Peach trees, Marburger Orchard, Fredericksburg, Texas (Photo by Andrew Edmonson)

Watering: It is important that stone fruits receive regular, deep watering. It is more important to water trees deeply than frequently. Watering once a week is sufficient in the cooler months and twice a week in the summer. Shallow watering will produce shallow roots. Deep watering is best achieved with drip or bubbler irrigation.

Fertilizing: In general, stone fruits do not require highly fertile soils for good production. Soils rich in nitrogen encourage leaf growth rather than flower and fruit production. Trees should be fertilized with a topdressing of compost during the dormant season. If trees are fertilized during flowering or fruiting periods, they are more likely to produce leaves than fruit.

It is wise to have a soil test done before planting to test for phosphorus and potassium levels and micronutrients. Phosphorus is necessary for genetic formation (plant DNA and RNA molecules are linked by phosphorus bonds) and stimulation of fruit, flower, and root production. Phosphorus can be supplemented with bone meal or granular phosphate if needed. Bone meal is a slow-release supplement, so it is good to apply it early when the tree is planted and again every fall. It can be mixed in with the soil when the tree is planted at a rate of ½ cup per ½ inch of trunk diameter. In consecutive years, simply mix the amendment into the top layer of the soil around the base of the tree. Keep in mind that an overabundance of phosphorus can tie up iron and zinc, especially in high-pH soils.

Potassium deficiency is not a common problem. If a problem is discovered, greensand is an organic source. There are a number of good products that supply micronutrients. Foliar sprays of fish emulsion and seaweed should be part of a regular maintenance program.

Pruning: Fruit trees need to be pruned for maximum fruit production. The primary objective is to open the tree to light and air. Peach and nectarine trees are usually pruned using the "open" or "vase" method, in which large interior branches are removed over a period of a few years, leaving three or four scaffolding branches. Any branches that are less than a 45° angle from the trunk are good candidates for removal, because smaller angles will cause a weak point where branches may break. This type of pruning will also reduce brown rot pressure.

In the vase pruning method, the center of the canopy will be removed to allow more light and air circulation. Branch tips will be cut back as well to allow more terminal branching. Terminal pruning should be done just above a branching node, preferably at a node that faces the outside of the canopy, as this is the direction the branching will take. Keep in mind that multiple branches will emerge from the cut, and if the limbs are not pruned back far enough, there will be many tiny branchlets that will restrict light and air circulation.

Plum trees are commonly pruned using the "central leader" method. Apricot trees may be pruned in a "modified central leader" method in which four or fewer central leader branches are allowed to develop rather than one. In all of the pruning methods, terminal, crossing, and branches at less than 45° angles should be pruned in the same manner as in the open or vase method. Fruit trees can also be pruned into espaliers against trellises or walls, a dramatic option to which fruit trees lend themselves beautifully.

As recommended for any tree, dead or diseased wood should be removed immediately. Make sure the cut is well below any diseased portion of the tree, as lower portions may be infected without showing symptoms. It is also important to clean pruning tools before going on to the next tree. Rubbing alcohol wiped over the blades will prevent the spread of the disease from one plant to another. It is also a good idea to oil pruning shears at the end of the day to prevent rusting.

Thinning: Large, full fruits can be grown by thinning the fruits as they mature, allowing the tree to devote its energy to one fruit every 6 to 8 inches. This should be done no later than 30 days after bloom.

APRICOT
Prunus armeniaca

Description
Because apricots flower very early, the flowers are commonly stunted by late freezes and fruit production is sporadic in Texas. Varieties that bloom later in the season are preferable. There are better choices if space is limited; however, as a specimen, it is a really beautiful tree. The pink blossoms appear before the leaves in the spring, making a striking display. Apricots are self-pollinating so do not require two varieties for pollination, but pollination will be improved with another variety.

Varieties
Blenheim Medium-sized to large, yellow fruit with orange tint. Firm, juicy, pale orange flesh. Good for canning and drying. Ripens late June to early July. 400 to 500 chilling hours.

Bryan A variety discovered in Dublin, Texas. Medium-sized fruit with yellow-orange skin and orange flesh. Ripens early June. 600 to 700 chilling hours.

Vase pruning on peach tree

Chinese (**Mormon**) Tree is medium in size and spreading. Fruit small and clingstone. Skin color orange with a red blush. Cold and frost hardy; sets heavy crops of small to medium-sized sweet fruit. Recommended for areas prone to late-spring frost. Mid- to late harvest season. 700 chilling hours.

Early Golden Medium-sized to large, good-quality fruit. 450 chilling hours.

Gold-Kist Large, orange, freestone fruit. Requires fewer chilling hours so a good choice for South Texas gardens. Pollinates well with 'Katy.' 300 chilling hours.

Harglow A late-blooming variety with bright orange, firm, and flavorful freestone fruit. Ripens late July. 800 chilling hours.

Katy Large, all-purpose freestone fruits with sweet flavor. Ripens early June. 400 chilling hours.

Moorpark Large, oval, somewhat flattened fruit. Skin color orange and yellow. Large, vigorous tree. Ripens July to late August. 600 to 700 chilling hours.

Royal Large, oval, yellow-pink fruit. Good for fresh eating, canning, or preserving. Ripens in September. 400 to 500 chilling hours.

Tisdale Medium-sized orange fruit. Good producer for Texas. 600 chilling hours.

Tomcot Produces flowers over a 3-week period to accommodate snap freezes. Large, firm, orange fruit. Can be self-fertile but will appreciate another variety for increased pollination. Ripens early. 600 chilling hours.

Apricot fruit (Photo by Craig Ledbetter)

NECTARINE
Prunus spp.

Nectarines are non-fuzzy peaches. There are several varieties recommended for Texas:

Armking Medium-sized to large cling fruit. Ripens late May. 600 chilling hours.

Crimson Gold Medium-sized freestone fruit. Ripens mid-June. 450 chilling hours.

Red Globe Large freestone fruit. Ripens mid-July. 850 chilling hours.

Red Gold Small, yellow-fleshed, dark red clingstone fruit with some yellow on skin. Ripens late April. 850 chilling hours.

PEACH
Prunus persica

Description

Peach trees are truly beautiful, especially when they are in full bloom. The bright pink blossoms occur on the previous year's growth before the leaves appear, and the homegrown fuzzy, reddish-yellow fruits are even sweeter and juicier than store-bought fruit. They are self-fertile, so only one tree is required for fruit production; however, planting two varieties will encourage a more productive harvest. Standard-sized trees range 15 to 25 feet in height, semi-dwarf are usually less than 12 feet, and dwarf varieties grow to 9 feet. The standard varieties make beautiful accents in garden bed corners or along a walk as a tree allee or as architectural accents in a formal design.

Dwarf varieties produce full-sized fruit and can be grown in containers. All sizes can be espaliered against a wall or trellis.

Varieties

There are quite a few varieties of peach trees for Texas. There are many considerations in selecting the best variety: yellow or white flesh; round or doughnut shape; freestone or clingstone; standard, semi-dwarf, or dwarf. The most important consideration is the number of required chilling hours for fruit to set. Late cold snaps can ruin developing blossoms. The later the tree blooms, the better chance it has to hold the blooms and develop fruit. Following are some proven varieties for Texas with a brief description and number of required chilling hours. It is important to remember that urban areas are typically warmer and therefore have fewer chilling areas than rural areas. Check local conditions before selecting varieties.

Belle of Georgia Freestone fruit with white flesh. Very sweet, low acid. Ripens end of July. Generally self-fertile but appreciates another variety for pollination. 850 chilling hours.

Bounty Large, reddish-colored freestone fruit with yellow flesh. Ripens early July. 800 chilling hours.

Denman Large, yellow-fleshed freestone fruit. Ripens mid-July. 850 chilling hours.

Dixieland Medium-sized freestone fruit with yellow flesh. Ripens early July. 750 chilling hours.

Peach tree, Marburger Orchard, Fredericksburg, Texas

Earligrande Yellow semi-cling fruit. Produces very early, usually in early May in South Texas. 275 chilling hours.

Elberta Red-yellow freestone fruit. Ripens late July. 850 chilling hours.

Floridaking Medium-sized semi-freestone fruit with yellow, firm flesh. Ripens mid-May. 450 chilling hours.

Harvester Medium-sized to large freestone fruit with yellow flesh. Resistant to bacterial leaf spot. Ripens late July. 750 chilling hours.

Indian Cling White-fleshed clingstone fruit. Ripens end of July. 750 chilling hours.

June Gold Large, high-quality, yellow clingstone fruit. Ripens late May to early June. 650 chilling hours.

Juneprince Yellow, semi-cling fruit. Ripens mid-June. 650 chilling hours.

La Feliciana Medium-sized to large, yellow freestone fruit. Heavy producer. Ripens mid-June. 450 to 550 chilling hours.

Loring Large, firm, yellow-red freestone fruit. Disease resistant. Ripens early to mid-July. 750 to 800 chilling hours.

Monroe Large, yellow-fleshed freestone fruit. Very good flavor. Ripens mid-August. 850 chilling hours.

Ranger Large, reddish-yellow fruit produced on vigorous trees with dark green foliage. A freestone variety with firm flesh that holds up well after harvest. Ripens mid-June. 900 chilling hours.

Redglobe Large, red freestone fruit with yellow flesh. Ripens late July. 850 chilling hours.

Redhaven Medium-sized to large freestone fruit with yellow flesh. Heavy bearing and resistant to leaf spot. Ripens early July. 950 chilling hours.

Redskin Large freestone fruit with yellow flesh. Ripens end of July. 750 chilling hours.

Rio Grande Medium-sized to large, yellow semi-freestone fruit with a red tint. Very sweet tasting, low acid. Ripens early May. 450 chilling hours.

Saturn Medium-sized to large, freestone fruit with gold skin. Double-pink blossoms with pink centers. Fruit appears flattened in a "doughnut" shape. A very interesting specimen. Ripens late July. 550 chilling hours.

Texroyale Yellow-fleshed freestone fruit. Ripens early June. 600 chilling hours.

Texstar Yellow semi-cling fruit. Heavy producer. Ripens early June. 450 chilling hours.

Tropic Snow Medium-sized, white-fleshed freestone fruit. Grows well in southern Texas. 200 chilling hours.

PLUM
Prunus spp.

Description

Many varieties of plums do well in Texas. They are very striking in the landscape, with fruit in various shades of red, yellow, blue, and purple and foliage colors with varying hues of green, red, and purple. The pulp is generally orange. The pink or white flowers appear shortly before the foliage in the spring.

Most plum varieties are not self-pollinating, requiring two different varieties for pollination. There are exceptions that will produce fruit without a pollenizer; however, it is always best to plant more than one variety for optimal fruit production. Standard-sized plum trees should be spaced 15 to 20 feet apart. Dwarf varieties can be spaced as close as 10 feet apart. It generally takes about 4 years for the trees to begin producing fruit.

Two species of plum trees are generally grown for fruit production: *P. salycina* (Japanese plum) and *P. domestica* (European plum). The European varieties are generally sweeter and drier with a firmer flesh and are commonly used for prune production. They usually have higher chilling requirements and a greater susceptibility to fungal diseases. The Japanese varieties are not usually found in supermarkets because they do not travel well. The wonderfully sweet, juicy flesh bruises easily in transport. Most of the varieties planted in Texas are crosses between European and Japanese varieties.

Varieties

Allred Red leaves and red fruit. Usually self-fertile. Good eating quality. Ripens in June. 650 chilling hours.

Gulf Ruby Yellow flesh and red skin. Requires pollenizer. Ripens early. A good variety for South Texas. 250 chilling hours.

Plum tree (Photo by Forest Wander)

Gulfgold Firm, yellow flesh. Requires pollenizer. Ripens early so a good choice for southern counties. 350 chilling hours.

Methley Medium-sized to large fruit with purplish skin. Amber to reddish pulp. Great pollinator. Ripens early June. 450 chilling hours.

Morris Large, round fruit with purplish skin and red flesh. Resistance to fungal problems. Ripens early June. 800 chilling hours.

Ozark Premier Japanese variety. Large fruit with bright red skin and yellow flesh. Self-fertile. Ripens late June. 800 chilling hours.

Santa Rosa Japanese variety. A large purplish plum with amber flesh. Self-fertile. Ripens late June. 700 chilling hours.

Stanley European variety. Oval shaped with deep purple flesh and golden-yellow freestone pulp. Self-fertile. Ripens early September. 800 chilling hours.

Wickson Japanese variety. Large, heart-shaped, greenish-yellow fruit. Very sweet, translucent flesh. Little or no tartness. Partly self-fertile. Ripens in July. 500 chilling hours.

 # Shrubs

BLUEBERRY
Vaccinium ashei

❋ **DESIGN USES:** Blueberry plants make wonderful hedges and screens, although they are deciduous. White to pink flowers in the spring and vibrant red foliage in the fall, with clusters of green to pink to blue berries in the summer, make these plants a multiseason design element. They require acidic soil, so most Texas gardeners will need to grow blueberries in containers. Containerized plants can be used as accents set in beds of colorful plantings.

Description
The blueberry bushes that grow in Texas reach about 10 feet in height and are deciduous. The white to pink spring flowers and reddish fall foliage, as well as the delicious berry fruit that ripens through varying shades of green, red, purple, and blue, make this shrub a worthwhile effort.

Locate blueberry plants where they will receive at least 8 to 10 hours of unfiltered sunlight per day. Because of their height, blueberry shrubs should be located toward the back of the bed, taking care not to shade other shorter plants. Container-grown plants can be set in the garden and do not need to be brought inside during the winter months. Containers should be well draining and 20 to 30 gallons in size. Half whiskey barrels are a good choice. Just be sure the containers have adequate drain holes drilled in the bottom.

Cultivation
Blueberries require high-acid soils with plenty of organic material and water. They will grow in soils

Blueberry plants at Chicamaw Farm, Bastrop, Texas (Photo by Andrew Edmonson)

with a pH below 5.5, and that is a very limited region in Texas, mainly some areas of East Texas. Otherwise, would-be blueberry growers will have to resort to container growing. For in-ground planting, soil pH can be lowered slightly by adding sulfur; however, sulfur is slow acting and will not amend most Texas gardens sufficiently.

Planting: Blueberry plants are shallow rooted, with the majority of their roots in the top 10 inches of soil. In their native environment, blueberries grow at the edge of bogs above the water table. They do best when these conditions are duplicated as closely as possible, with thick layers of acidic mulch. During the first year when the plants are small, a couple inches of mulch will be sufficient. After the first year, mulch levels should be 6 inches in depth. Pine straw is a good mulch choice.

For in-ground planting, prepare raised beds prior to planting with a 5:4:1 ratio of pumice, peat moss, and loam. Coconut coir is a more sustainable alternative to peat moss. Well-decomposed pine straw and pine bark are good high-acid additions as well. Container-grown plants should be planted in a mixture of pine bark and peat moss or coconut coir. Be careful that the soil media is well moistened and does not dry out. It is very difficult to rewet once it has become dry.

Watering: Although blueberry plants will not survive in standing water, they require quite a bit of irrigation. In fact, drip systems generally do not provide sufficient water. Micro emitter irrigation systems are a better choice for blueberries, as they put out five to ten times more water per hour. The plants should be watered two to three times per week, and amount of water should be adjusted with the season, plant size and, soil texture. Emitters should be spaced away from the trunk so a wider radius of the root zone will be saturated. As a rule, the root zone will spread out about as far as the canopy of the shrub. Water deeply and allow the soil to drain before the next watering. Water sources with high salt content are generally high in alkalinity, and blueberries will not tolerate this condition. Harvested rainwater or water from air-conditioner condensate will generally provide lower-alkaline sources.

Fertilizing: Blood meal, cottonseed meal, feather meal, soybean meal, and alfalfa meal are good nitrogen sources, and kelp meal will provide trace minerals. Blueberries respond well to foliar feedings of fish emulsion.

Pruning: Blueberry bushes do not require heavy pruning. Dead, diseased, or crossing branches should be removed, as well as tiny twigs. Heavy tip pruning is discouraged because fruit forms on the tips. A little tip pruning will encourage more prolific fruit production on the remaining branches. On mature plants, prune lower branches by 20 percent to encourage fruit production and to prevent the fruit from coming into contact with the soil.

Harvesting: Wait until blueberries are fully mature before harvesting. Unripe berries are bitter and do not ripen well off the tree. Berries should be harvested every 5 days or so. They continue to mature after they have turned blue and will continue to plump as they mature. Ripe berries will come off easily with a gentle tug. It is best to harvest during dry weather.

Controlling pests: Blueberries are not troubled by many pests, except birds. Birds can wipe a blueberry bush clean before the gardener ever gets a taste. They will even eat the unripe berries. Because the wood of the blueberry is brittle, birds can break off branches in their feeding frenzy, especially if they are trapped inside bird netting, so it may be necessary to build a frame to drape the netting over and anchor the netting to the ground.

Varieties

There are two blueberry varieties cultivated in the United States, highbush and rabbiteye. Rabbiteye are the preferred varieties for Texas. They are better adapted to high temperatures and humidity. 'Tifblue' has been shown to be self-fertile but will perform better when planted with another variety. Different blueberry varieties produce at different times. The season can be extended by planting varieties that produce at varying times. For best results, plant three varieties for good cross-pollination. Blueberry plants are pollinated by bees, so it is also important that there is a local bee colony, and nonspecific insecticides such as pyrethrum should not be used in the area.

Alapaha A strong variety with medium-sized fruit. 500 chilling hours.

Austin Good producer of medium-sized berries in midseason. Good pollenizer. 500 chilling hours.

Brightwell Vigorous, upright growth. Produces heavy yields of medium-sized fruits in midseason. 400 chilling hours.

Climax An early-ripening variety over a short time period. Production is moderate to high. 450 chilling hours.

Ochlockonee Vigorous producer of medium-sized to large fruits. 700 chilling hours.

Powderblue Good production of medium-sized fruit late in the season. 600 chilling hours.

Premier Medium-sized to large fruits. Good pollenizer. 550 chilling hours.

Tifblue The most widely planted and best producer for Texas. A heavy producer of large, light blue berries late in the season. 650 chilling hours.

Vernon Vigorous grower that produces in mid- to late season. 550 chilling hours.

Woodard Moderately heavy producer of large, good-quality fruit. Ripens early. 350 chilling hours.

Chicamaw Farm

Bill McCranie and his wife, Nancy; along with their children, Cassady and Patrick; a number of Catahoula leopard dogs; and two dachshunds own a pick-your-own blueberry operation in McDade, Texas. McDade is about 30 miles east of Austin, well out of the normal blueberry-growing range. Bill has been involved in organic methods for over a decade and has included biodynamic cultivation principles for 8 to 9 years.

Rudolph Steiner first introduced biodynamic farming principles in the 1920s. The practices have some things in common with organic cultivation, viewing cultivation as an interrelated system of plants, animals, and microbial activity, as well as the human and spiritual elements, looking at the end-user connection. Steiner proposed his cultivation principles to be an experiment as an extension of his philosophical, spiritual, and holistic theories that certain biological preparations would have a positive effect on soil and plant health. Like organic principles, biodynamic cultivation also includes the use of manure, involving more specific preparations, burying the manure in the ground packed into cow horns, and using the resulting humus in bed preparation. Specific herbal preparations added to compost heaps are also a part of these cultivation practices. There is more than a little question about how these preparations affect the overall health of plants and the soil where they grow. It is amazing to be growing blueberries in McDade, Texas, though.

ELDERBERRY
Sambucus spp.

❀ **DESIGN USES:** Elderberries are deciduous shrubs with compound leaves that give them an airy effect. They are one of the few edible plants that do well with limited sunlight, so they can be planted as an understory shrub on the edge of the overhead tree canopy. Yellow- and purple-leaved varieties add special interest to otherwise shaded areas, although the use of these plants may be limited to northern and eastern counties of Texas. Sprays of elderberry flower clusters vary in color from white to pink and are quite showy, as are the berry clusters.

Elderberry flowers (Photo by Sebastian Maćkiewicz)

Description

The elderberry is a multitrunked, deciduous shrub that grows to a height of around 8 to 12 feet. New growth occurs in the form of canes growing from the base of the trunk, as well as on the previous year's growth. Elderberries are more common in north-central and East Texas, and it is important to check the suitability for the zone when planting. Some varieties will only grow as far south as zone 7. Elderberries tend to spread by underground runners, creating thickets in the wild. They can be kept in bounds by pulling up the runners and clipping them at the base of the mother plant. Some elderberries are self-fertile. If there is room, it is best to plant two or more varieties for increased production.

In midsummer elderberry shrubs produce a profusion of small white or pink flowers clustered in corymbs. These flowers become edible black berries in the late summer and early fall and should be allowed to ripen on the plant. The unripened berries and all of the rest of the plant contain a low quantity of cyanide, so be cautious to harvest only the ripe berries. The berries are a favorite of birds, but there are usually enough to share. Elderberries are very high in vitamin C. They are mostly used in baked goods or for making jellies and wine and have been grown historically for various medicinal uses.

Cultivation

Elderberries are a great choice for Texas and are not prone to disease or insect problems.

Planting: Elderberries are one of the few edible plants that can take some shade and can tolerate wetter soils (but not standing water). Where they are naturalized in the wild, they are usually found growing in an understory condition, on the edge of woodlands, where their roots can reach into deep, moist soil and their branches can reach out to the light. Elderberries can be planted in moist, well-drained soil any time of year, although fall is always the best time of year to plant hardy perennials.

Fertilizing: Like any cultivated plant, elderberries will appreciate a topdressing of well-decomposed compost at the beginning of the growing season. They particularly like rich soil, although they grow in the wild without any additional fertilizer.

Pruning: Prune in the early spring, before the leaves bud. Prune back all but five or six of the most vigorous canes produced in the previous year's growth and one or two of the 2-year-old canes. Trim any crossing or internal branching and remove any dead wood.

Varieties

Elderberries in the species *canadensis* are prolific bearers and generally fruit earlier than other varieties.

Aurea Golden-yellow foliage that brightens an understory area. Needs at least 6 hours of sun per day or it will lose its golden color. Pollinates with 'Adam's,' 'John's,' or 'Nova.'

Adam's A heavy producer with very large flower heads. Produces late in the season. 'John's' is a good pollinator.

John's Large berry clusters. Cross-pollinates with 'Nova' or 'York.'

Nova Produces large fruit and has a high yield. Large white flower clusters. Self-fertile.

York Very good producer of large berries. Very similar to 'Nova,' except fruit ripens 2 weeks earlier.

Elderberries in the species *nigra* have some very appealing color and texture interests. They are better suited for the more northern and eastern counties of Texas. Check suitability for planting zone.

Black Beauty Bold, dark purple-red leaves that contrast beautifully with the pink 10-inch flower clusters. Needs plenty of sun to retain foliage color.

Black Lace Deeply lobed variety with a texture similar to that of a Japanese maple. Deep burgundy foliage and pink flowers.

Cut Leaf A 6- to 8-foot shrub with deeply lobed leaves. A very striking accent or hedge.

Goldbeere Upright growth habit with golden berries against light green foliage.

Haschberg Vigorous spreading growth habit, to 10 feet in height. Large fruit clusters.

Samdal Large fruit clusters. Ripens late summer.

Variegated Grows to 8 feet. Creamy variegation on bright green leaves. Fruits in September.

FIG
Ficus carica

❋ **DESIGN USES:** Figs can be used as a large, dense shrub, or the lower vegetation can be removed to create a multitrunked tree. The extremely bold-textured foliage makes a dramatic, semitropical statement in the garden. Figs lose their leaves in the wintertime, and the remaining branches are somewhat architectural. Be sure to plant colorful and evergreen winter plantings in the same area for winter interest. Also use figs on overhead arbors and for climbing and hiding places in children's gardens.

Description

Figs are wonderful selections for Texas gardens. They are usually multitrunked and generally reach between 20 and 30 feet in height. There are a wide variety of cultivars with a rich array of fruit colors, from rust to brown, purple, and black. Most common fig varieties are well adapted to the warmer regions of Texas. Some can be damaged in severe freezes, but the roots are usually spared and the plants will generally come back from the base of the tree even if the aboveground growth is damaged. The leaves of the fig have three or five lobes and are very large and bold in texture, making a dramatic addition to the garden.

The fruit of the fig is unusual. It is formed from male and female flower parts enclosed in the stem. The fruit evolves from stem tissue rather than a mature ovary. In some varieties, there are many tiny seedlike structures in the fruit that are unfertilized ovaries. Fruit will change color as it ripens and is best if allowed to ripen on the tree. Figs with open "eyes" may be susceptible to the dried fruit beetle feeding on the fruit or souring if rain enters the fruit.

There are four types of figs that have been classified according to their method of fruit production. The caprifig produces small, inedible fruit. It is grown for its pollen, which is used as a pollinator for two other types of fig, the Smyrna and San Pedro. The Smyrna fig produces a large fruit with true seeds. These figs are commonly grown to make dried fruit. The San Pedro fig can produce two crops per year. The first crop is produced on the previous year's growth and does not require pollination. The second crop is produced on new growth and requires the caprifig for pollination. Both the Smyrna and San Pedro fig types need the fig wasp (*Blastophaga psenes*) for pollination.

Only the fourth type of fig, the common fig, is recommended for Texas, and all of the recommended species in this section are of this type. The common fig does not require a pollenizer; it is parthenocarpic, which means the fruit is produced without fertilization.

Cultivation

Like most fruit trees, figs need to be planted in a well-drained, sunny location to do well. They do not require an excess of fertilizer or water to produce an abundance of sweet, edible fruit; however, they respond well to a yearly topdressing of compost. Compost should be applied after figs have gone dormant or during the growing season. If they are top-dressed in the fall, new, tender growth could be harmed by freezing weather. Unlike some fruit trees that require thinning of branches and fruit for good production, figs do not need much pruning. Severe pruning will limit the amount of fruit produced.

Sanitation is very important in fig cultivation. Fallen fruit and leaves around the base of the plant should be removed to prevent fungal disease from occurring. Figs can be susceptible to fungal diseases, especially if the soil does not drain properly or water is left on the leaves during the night. It is best to irrigate at the base of the trunk with a drip system if possible. If overhead spray systems are used, water in the early morning so the leaves will have time to dry completely during the day. If fungal problems do occur, remove diseased leaves and treat with an organic fungal treatment.

Bold texture of fig leaves

Varieties

Alma A late-season variety that produces good-quality fruit with a sweet flavor. Produces at a younger age than other varieties. Good resistance to fungal diseases. Susceptible to freezes if planted north of Austin. Closed eye.

Brown Turkey Medium-sized to large reddish-brown fruit tinged with purple and has reddish-pink pulp. Can have a problem with "souring" and more winter sensitive than 'Celeste.' Fruits on new growth from the base if top growth killed by freezing weather. Moderately closed eye.

Celeste Small, brown, purplish fruit with a pinkish pulp that is rich and sweet. The most cold hardy. Closed eye.

Osborne Prolific Medium-sized fig with bronze skin and amber to light pink pulp with a sweet, rich flavor. Some cold tolerance. Partially closed eye.

St. Anthony Marseilles Large, yellow fruits produced on the previous season's growth. Closed eye.

Texas Blue Giant Large, purple-skinned fruit with pinkish-amber flesh. Closed eye.

Violet de Bordeaux A dwarf fig tree that produces dark purplish fruit with strawberry-colored pulp. Very good choice for container planting. Closed eye.

☀ Perennials

During the design phase, it is important to locate perennial plants in well-prepared permanent beds. Along with the edible trees and shrubs, perennial plants will form the framework of the design. Many herbs are perennial as well and should be included in this framework. It is important to remove all weeds and prepare the beds with plenty of organic matter and soil amendments such as compost. The soil should also be loose and well drained. Various meal fertilizers, earthworm castings, lava sand, mycorrhizal fungi, and humate will

Blackberry blossoms (Photo by James Beesley)

also help build the soil in these permanent beds. If the soil is heavy clay, work in sulfur and sand as well.

ARTICHOKE
Cynara scolymus

☀ **DESIGN USES:** Artichokes are perhaps the most exotic plants in the edible landscape palette for winter gardens in Texas. The deeply lobed, gray-green foliage grows from the base of the plant, forming prehistoric-looking mounds of dramatic foliage with thistlelike globes growing on stalks from the center. Use as corner accents or as architectural geometry in modern gardens.

Description
Even if artichoke plants did not produce a wonderful vegetable, they would be prized in the garden for the dramatic silvery foliage and vivid purple, thistlelike flower head. Artichoke plants begin producing flower heads in the second year, which are the edible portion of the plant and should be harvested before the flowers are fully formed. The more tender "heart" is the fleshy base to which the bracts are attached. The bracts are the fleshy, point-tipped part that is delicious dipped in melt-

Artichoke foliage (Photo by James Beesley)

ed butter. Artichokes grow from 3 to 6 feet tall and wide and add a beautifully dramatic accent to the winter garden. They can be sensitive to severe winter temperatures and will not usually survive temperatures below 15°F.

Cultivation

Artichokes require well-drained soils and will appreciate well-prepared beds with plenty of decomposed organic matter. They grow during cooler months and, like many Texans, go dormant in the summer heat. It is important to apply mulch to protect the plant's roots into the following year. New plants can be started from the little plantlets that grow at the base of the parent plant. They can also be started from seeds, although seeds may not produce the same plant as the parent plant. Crown divisions are more reliable.

Artichoke flower (Photo by Andrew Edmonson)

Varieties

Green Globe and **Green Globe Improved**
Green, globe-shaped fruit. Improved variety is cultivated to have fewer spines and larger heads.

Purple Italian Globe Produces large, flavorful heads. More tolerant to heat and cold.

ASPARAGUS
Asparagus officinalis

❋ **DESIGN USES:** The light, airy foliage of asparagus creates a wonderful background in the edible garden. The airy leaf texture is a nice contrast to more densely textured plants in the same bed. The plants also make a pretty background to brightly colored foreground plants. The female plants have red berries and are more decorative but are not as productive as the male plants.

Description
The edible portion of the asparagus plant is a spear that grows from the rhizomatous roots. The beautiful, 4- to 5-foot-tall, feathery foliage is strictly a bonus.

Cultivation
Asparagus plants can live more than 20 years, producing delicious edible spears in greater abundance each season once they become established, so it is very important to prepare the asparagus beds correctly when planting the crowns. Select a permanent spot where the asparagus will grow. Because of the plant's height, it is a good idea to

Asparagus foliage, Natural Gardener, Austin, Texas

lay out asparagus beds where they will not shade shorter plantings. Asparagus plants require at least 4 to 5 hours of direct sunlight per day to gather enough energy to produce a good stand of spears. The crowns should be spaced at least 18 inches apart in a well-prepared bed. Asparagus spears may be more spindly in areas south of San Antonio. Warm nights that are so common in the southern parts of Texas cause the photosynthetic energy to be used for current growth rather than stored in the roots. If soil tests show a pH less than 6.0, which can be the case in areas of East Texas, additional lime may be needed as a soil additive at an application rate of 2 to 6 pounds per 100 square feet.

Planting: Asparagus plants are usually grown from 1- or 2-year-old crowns, although they can be grown from seeds. Growing from seeds will add a year to the time before the spears can be harvested. Even after the crowns are planted, full harvesting should not be done for at least 2 years. Seeding would make the wait 3 years, and that is a long time to wait for this tasty spring treat. Asparagus crowns should be spaced 15 to 18 inches apart on center. A good method for planting is to dig a trench, or individual holes, depending on straight or triangular spacing, 12 inches deep and wide. Layer 2 inches of soil, 4 inches of well-decomposed compost, and a good handful or two of bone meal or other organic phosphate in the hole and repeat with a second layer. The hole will not be filled at this point. Mound up the soil mixture to form a base for the roots of the crowns, and spread out the roots down the side of the mounds. Cover the roots with 2 to 3 inches of compost and soil mixture, and water well to settle the roots. It is good to water the crowns in with a fish emulsion and seaweed mixture to give them a boost. As the spears grow, continue layering a mixture of compost and soil over the roots every week or two until the trench (or hole) is filled. When the trench is completely filled, spread 2 inches of hardwood bark mulch or pine straw on the soil surface.

Fertilizing: Asparagus is a heavy feeder. During the first year, the plants should be able to draw on the reserves of compost and soil amendments applied during the bed preparation. A topdressing of compost should be applied yearly after the spears are harvested.

Pruning: It is best to leave the fernlike top growth of the asparagus alone. The plants will die back to the ground after the first freeze. This is a good time to add a layer of mulch. In areas where temperatures do not generally reach the freezing mark, remove the foliage after it yellows.

Harvesting: Harvest the spears sparingly in the first 2 years. While it is tempting to harvest all of the spears, they will produce heavier future growth if some are left to grow. Even in the third year and beyond, it is good to leave a few spears to produce new foliage. Harvest the spears when they are 6 to 8 inches long and about ½ inch in diameter. The tips should be tightly closed. Make a diagonal cut right above ground level. Stop harvesting when the spears are ³/₈ inch in diameter, and allow them to grow for the next year's production.

Varieties

There are three main types of asparagus in cultivation. The first includes the old standards 'Mary Washington' and 'Martha Washington.' These varieties include female plants and are not as productive as the newer cultivars. The second is the Jersey series. It includes the varieties 'Jersey Giant,' 'Jersey Knight,' and 'Jersey Supreme,' which have been hybridized for more northern climates. They are all male, very productive, and more adapted to northern areas of Texas. The varieties developed by the University of California, or the UC series, have been chosen for their ability to grow in hotter climates and should do well in southern parts of the state.

Apollo, Atlas, and **Grande** Varieties developed by crossing a Jersey variety and a UC 157, which combines the productivity of the Jersey variety and the heat tolerance of the UC variety. Produce both female and male plants. Female plants can be removed and replaced with male plants for greater productivity if desired. Resistance to fusarium wilt and rust has been shown, although these are relatively new varieties, so more time may be needed to determine ultimate resistance to these diseases.

Jersey Giant Male variety with good yield. Susceptible to fusarium wilt.

Jersey Knight Male variety with thick spears. Good resistance to fusarium wilt.

Jersey Supreme Male variety with good production.

Purple Passion All-purple spears that turn green when cooked. A striking addition to the garden. Plants not prolific, but spears and plants larger. Sweet and tender.

UC 157 Good yield but smaller, thinner spears than Jersey types. Resistant to fusarium wilt and rust. Heat and drought tolerant.

UC 72 Developed from the 'Mary Washington' variety. Good yield of green spears. Resistant to fusarium wilt and rust. Heat and drought tolerant.

Bramble Fruits

BLACKBERRY
Rubus spp.

❋ **DESIGN USES:** Blackberry vines can be used as shrubs or vines depending on variety. The thorny varieties are useful to keep out would-be intruders on fence lines. Grow vining types on trellises or as hedgerows on split-rail or other short fences. Make sure to leave room to maintain and prune out old canes. Shrub types can be grown as loose groupings or as accents in garden beds. Blackberry vines are very pretty as backdrops to other plants in the garden. With delicate white to light pink flower bouquets in the late spring to early summer and varying shades of red to deep purple to black berry clusters in the summer, these plants are a dramatic addition to edible landscape design.

Blackberry flowers (Photo by Andrew Edmonson)

Description

Blackberries are very well suited for Texas. They are prolific producers and relatively care-free. Blackberries are either upright or erect types that do not require trellising or trailing varieties that do need to be trellised. The perennial, bramble-type vines form large arching shrubs if they are not trellised. Both the erect and trailing types of blackberries can be further divided into thorned or thornless varieties.

Blackberry canes grow from the base of the plant. The first canes to emerge are called primacanes. These primacanes become more woody in the second year, when they are called floricanes. As the name implies, these canes flower and produce fruit. The floricanes are removed after the fruit is harvested, to leave the newly emerged primacanes for the following year's fruit production.

Blackberry fruits are not true berries but a collection of small fruits, or aggregate fruit. The fruits are whitish when they emerge, and they turn red and then black as they mature. The mature fruits are sweet and juicy and should be picked when they are black and come off easily from the vine. They will become overripe and mushy if left on the vine too long, so it is important to harvest every few days during the 4 to 6 weeks of harvest.

Cultivation

Blackberry plants like well-drained, sandy soil, with a pH range of 4.5 to 7.5. Since they will be planted in a more permanent location than many garden plants, it is important to make sure the soil is well prepared before planting. Adjust soil as needed to balance pH, and apply soil additives, as well as organic matter according to the recommendations in the soil preparation section of this book. It is also important to be sure all existing weeds and grass have been removed from the beds, as it will be difficult to remove them after the canes have matured.

Planting: Blackberry vines should be planted at the same depth that they were grown at the nursery. The vines can continue to produce for 15 years, but they are most productive in the third through the eighth years, so it may be a good plan to replant every 5 to 8 years and retire the old plants as they fail to produce.

Fertilizing: Blackberries do not require an abundance of fertilizer; simply top-dress with well-decomposed compost before bloom and in the fall. Blackberries are susceptible to iron chlorosis in high-alkaline soils. Chlorosis will appear as a yellowing between the leaf veins, especially in new growth. Iron chlorosis can be treated by adding chelated iron to the soil or applying iron as a foliar treatment. This is really just putting a patch on the problem though. It is better to amend the soil with sulfur and compost. If iron is applied as a foliar treatment, be careful not to spray walls or paving surfaces, as the iron will stain. It is also important to spray when temperatures are below 80°F and to coat all leaf surfaces completely.

Pruning: To increase fruit production, thin the floricanes to four to six strong canes before they flower. Canes can be kept at a length of 6 feet to encourage lateral branching and make harvesting easier. Prune the canes that produced in the current year to the ground after the fruit is harvested.

Controlling insects and diseases: Although blackberry plants are fairly easy to grow in Texas, it is important to maintain good sanitation practices to prevent disease. Cleaning up plant debris and maintaining good air circulation as well as planting in loose, well-drained soil should prevent fungal problems. It is also important to remove year-old canes after they have produced fruit to prevent insect infestation and diseases.

Varieties

Apache An almost thornless variety newly released from the University of Arkansas. Canes vigorous and erect and produce large, sweet berries. Fruit in mid-June.

Arapaho Erect, thornless variety. An early producer of medium-sized, flavorful fruit. A promising new variety, showing early signs of rust resistance.

Brazos A longtime standard in Texas. Erect plants and reliable producers. Fruit somewhat acidic, making it a better choice for cooking than eating.

Brison Erect plant recommended for more southern areas of Texas and in heavier soils. Large fruit.

Choctaw Erect canes that produce medium-sized to large, soft fruit.

Kiowa Early producer of very large, good-quality fruit.

Navaho Erect, thornless variety with firm, medium-sized, sweet fruit.

Rosborough Erect, vigorous grower. Fruit slightly larger and sweeter than 'Brazos.' Recommended for more northern and eastern counties in Texas.

Shawnee Erect canes that produce large fruit later in the season.

RASPBERRY
Rubus spp.

Description

Raspberry fruit can be red, black, or yellow. The fruit is different from blackberries, in that the core of the fruit pulls away from the outer part of the berry when picked. This is a sure way to discern raspberries from blackberries. Because of alkaline soils, raspberries are not as well suited to all areas of Texas as blackberries, although there are a few varieties that might be tried in areas with higher soil pH. It is probably best to try a few plants and see how they perform before planting a whole fence line. Counties farther north and east have a better chance with raspberries.

Cultivation

Raspberries like well-drained soil that has a pH between 5.8 and 6.8. They also prefer a little afternoon shade from the hot Texas sun. Heavy clay soils will need to be amended with sand and organic matter to make them looser.

Planting: Plant raspberries at the same depth they grew at the nursery, and mulch well with 3 to 4 inches of finely shredded hardwood bark mulch or pine straw.

Pruning: Raspberries are pruned in specific ways to encourage them to fruit at different times. Certain varieties are grown for spring to summer bearing (spring-bearing), and others are grown to bear in the summer and fall (ever-bearing). The spring fruiting varieties are pruned like blackberries, allowing the primacanes to mature into floricanes and then pruning them to the ground after the fruit is harvested. Ever-bearing raspberry cultivars produce fruit on the first-year primacanes, and the plants are cut to the ground in the late winter or early spring.

Varieties

Amity Large, dark red berries with good flavor. If pruned as spring bearers, they produce in the spring on 2-year-old wood and again in the fall on the first-year growth. Produce a single heavier crop if pruned to the ground in the early spring. Resistant to root rot and aphid infestations.

Autumn Bliss Ever-bearing variety similar in habit and production to 'Heritage.' Red berries produced in summer and fall. Good disease resistance.

Bababerry Very large, sweet, red fruit. Bears a large crop in summer, followed by a small crop in the fall.

Caroline Everbearing Large, cone-shaped, red fruit. A summer producer. Some resistance to root rot and yellow rust.

Cumberland Black Trailing variety with black berries. Good-flavored berries and vigorous growth.

Dorma Red Trailing variety produces red berries spring to summer.

Heritage Red Ever-bearing variety cultivated for spring and fall fruit. Good-quality red fruit. Does not need staking if cut to the ground in the spring.

Nova An almost thornless variety that produces medium-sized to large, red berries. Some disease resistance.

Willamette Spring-bearing variety. Large, round, dark red berries with a slightly tart flavor. Good disease resistance.

Dewberry flowers (Photo by Walter Siegmund)

SOUTHERN DEWBERRY
Rubus trivialis

Description

Dewberries are very hardy and grow wild in hedgerows throughout Texas. Blackberries are cultivated varieties of this wild cousin. The main difference between the two is their growth habit. Dewberry vines are shorter (about 3 feet) and more sprawling than blackberry vines. Dewberries are also more assertive and can easily take over a bed with very aggressive root systems. Be sure to give these vines plenty of room, and prune and dig vines that invade unwanted bed areas. Gloves are required when pruning or harvesting dewberries, as the vines are covered with tiny thorns. Trellising the vines will help keep the berries off the ground and make harvesting easier. Cultivation requirements and growing conditions are similar to those for blackberries.

Varieties

Austin Recommended for Texas. A vigorous producer of large, black berries. Begins flowering in March and forming fruit in late May.

Grapes
Vitis spp.

☀ **DESIGN USES:** Grapes are vines and need to be trained on a trellis. The yellow or reddish fall color of the bold-textured leaves, along with their dramatic clusters of pendulant fruit, give this

Grape vines, Natural Gardener, Austin, Texas (Photo by Andrew Edmonson)

plant a year-round appeal. Even when they have lost their leaves, the gnarled trunks of mature grapevines can be very interesting. Use rows of cordoned vines for an architectural element, grow on trellises as a backdrop, or allow them to hang from overhead arbors.

Description

Grapes are one of the oldest plants in cultivation, and their romance and allure are legendary. There is no more romantic gardening image than a lush grapevine dripping grapes from an overhead arbor. This image can become a reality for Texas gardeners with thoughtful selection of the proper variety, good sanitation, and diligent treatment for insect and disease problems. The High Plains of West Texas and areas of Central and North Texas with more arid climates are more favorable for grape growing than the Gulf Coast and East Texas regions where high humidity can cause a number of fungal problems.

The commonly grown table grapes in the grocery stores are the *vinifera* species. They are the introduced grape that the first American settlers brought over from Europe and the Mediterranean region for cultivation. Unfortunately, these introduced species are very susceptible to untreatable diseases in Texas, especially Pierce's disease and cotton root rot. Luckily, certain types of grapevines are native to Texas and do very well here. Years of selection and hybridization have produced some fairly decent grape varieties for Texas, especially for wine and jelly making, as they are generally more tart and tangy than the *vinifera* varieties.

Cultivation

The vines will need to be trellised, and the trellises should be in place before the vines are planted. Vines should be planted on 8-foot posts set 18 inches in the ground in concrete footings. Posts should be about 8 feet apart. The lateral branches will be trained on a trellis (espalier) system, with wires spaced at 42, 52, and 66 inches parallel to the ground.

Planting: It is important to plant grapevines in well-drained soil where they can get good air circulation. Grapevines will not grow well in heavy soils, and they should be planted in raised beds if local soils do not drain well. Grapevines do not like to compete for nutrients with other plants, so it is good to have mulch or a low-growing, nitrogen-fixing ground cover around the base of the vines.

It is best to plant grapevines in the winter when they are dormant. Plant as soon as they arrive from the nursery, or they can be "heeled" in beds for a day or two if necessary by covering the roots with soil. It is very important that the roots are not allowed to dry. Dig the holes for planting next to trellis posts. It is a good practice to have the holes prepared before uncovering the plant roots. When grapes are planted, the vines should be cut back to two buds on the strongest cane. Prune any other canes to the crown. Grape plants should be planted with the crown just above the soil. Covering the crown could cause future fungal problems. Fill the hole with existing soil, and firm the soil to prevent air pockets. Water the plants immediately after planting. Fish emulsion and seaweed mixed into the water will act as a plant stimulator.

Fertilizing: Grapes do not require any additional fertilization in the soil once they are planted. Additional soil amendments can cause new growth to be too lush and attract insects. It can also cause fungal problems at the ground level. Monthly foliar spraying during the growing season with fish emulsion and seaweed will help the plants stay healthy.

Some grape varieties are susceptible to iron chlorosis in high-pH soils. Chlorosis will appear as a yellowing between the leaf veins, especially in new growth. Iron chlorosis can be treated by adding chelated iron to the soil or applying iron as a foliar treatment. It is better to amend the soil with sulfur and compost. If iron is applied as a foliar treatment, be careful not to spray walls or paving surfaces, as the iron will stain. It is also important to spray when temperatures are below 80°F and to coat all leaf surfaces completely.

Pruning: Pruning should be done yearly in the late winter to encourage fresh growth and good fruiting in the following season. There are several pruning methods for grapevines. Generally 85 to 90 percent of the last year's growth will be pruned from the vines. As a rule, vines are pruned to just two buds in the second year, eight to twelve buds in the third year, and twenty to forty buds in the fourth year. Vigorous shoots should be ½ to 3/8 inch in diameter. If canes are less vigorous, the "cane system" of pruning should be used; if overly vigorous, the "single curtain" is preferable. Cordons are permanent "arms," while canes are selected each year.

Pruning Methods

Bilateral cordon system: This is the most common pruning method in Texas. Two lateral branches, one on each side of the main trunk, are selected to be trained on the 42-inch wire. Seven spurs are selected for each side and should have two buds each. Remaining spurs are pruned just above where they emerge from the cordon branch, leaving a stub with one node. This spur will become the next year's growth.

Cane system: Canes are produced when shoots mature and become woody. This is a good training method for less vigorous vines. All 1-year-old canes are removed, except one on either side of the main trunk. These two canes should be ½ to 3/8 inch in diameter and 4 to 6 feet in length, with six to twelve buds each. The canes are draped over the 52-inch wire. It is important to leave two short bud spurs to produce the canes for next year's production. Large-clustered varieties are usually pruned to the short cane pruning method, in which canes are pruned back to only one to six buds each.

Single curtain: This method is good for very vigorous vines. Two lateral branches are selected to be permanent branches tied to the top wire. Spurs are spaced 6 inches apart and have one to six buds each, depending on the vigor of the vine. More vigorous vines are allowed more buds.

Controlling pests and diseases: Grapes are highly prized by humans. They are also favorites of birds,

Cordoned grape vines, Natural Gardener, Austin, Texas (Photo by Andrew Edmonson)

as well as insect and disease predators. Specialized netting can save fruit from the birds. For insect and disease prevention, it is important to inspect grapevines almost daily during the growing season to catch problems before they get out of control. Grapes are notoriously susceptible to fungal diseases. Good sanitation practices will help lessen these problems. Remove any infected fruit, leaves, and vines to prevent the infection from spreading. Infected vines should be cut well below any signs of infection, and pruning shears should be disinfected with rubbing alcohol between cuts. Diseased debris should not be composted, as the entire compost pile could be infected. Any plant debris on the ground around the trunk of the vine should be cleaned up on a regular basis.

To help prevent fungal infections on the fruit, remove the leaves surrounding the clusters to increase air circulation. Before human-made chemical fungicides were used on grapevines, the vines were usually treated with Bordeaux mixture, a combination of copper sulfate and hydrated lime. It is considered an organic treatment, but there is some concern about copper pollution of waterways, and it can cause leaf burn in temperatures above 85°F. There are better organic choices available today.

Varieties

Pierce's disease is perhaps the most limiting factor for variety selection in Texas. There is no control for Pierce's disease, so the only way to prevent infection is to plant resistant varieties such as the following.

Blanc du Bois Grown for white wine production. Vigorous vine. Fruit ripens late June or July. Resistant to downy mildew but susceptible to black rot and anthracnose.

Champanel Mostly grown for jelly making. Small clusters of large, black grapes. Very acidic flavor until grapes have ripened completely. Susceptible to iron chlorosis if planted in soils with a high pH. Very resistant to fungal diseases. Very vigorous vine perfect for trellising on arbors.

Cynthiana Also known as 'Norton.' Cultivated by the renowned Texas viticulturist T. V. Munson from *V. aestivalis*. Has a milder flavor than many of the native varieties. Grown for the production of a dry wine. Fruit ripens late August to early September. Resistant to black rot and other fungal diseases.

Herbemont Small brownish-red fruit in medium-sized, compact clusters. The species *V. bourquiniana.*

Le Noir Also known as 'Black Spanish.' Grown for wine, juice, and jelly making. High in tannins and very acidic. Fruit ripens late July or early August. Susceptible to black rot and downy mildew.

Lomanto A T. V. Munson variety that produces medium-sized, black fruit on vigorous vines. Ripens early July.

Orlando Seedless Very vigorous vine that produces long, thin clusters of light green, seedless berries, which are grown for fresh eating. Very good flavor. Resistant to downy and powdery mildew but susceptible to black rot and anthracnose.

MUSCADINE GRAPES
Vitis rotundifolia

Muscadine grapes are well adapted to high heat and humidity and have a natural resistance to Pierce's disease, anthracnose, and black rot, making them a very good choice for Texas gardens. Modern varieties have larger fruit, are sweeter, and have thinner skins than native muscadines, which can be tough and bitter,

Muscadine grapevines require well-drained soil and full sun for plant health and maximum yield. They are native to the southeastern United States and prefer acidic soils, making them a good choice

Bird netting covering vines near Dripping Springs, Texas

for East Texas gardens. Muscadine grapes also have lower chilling-hour requirements for fruit production than many other *Vitis* species. Also, most muscadine varieties have pistillate flowers, requiring a pollinator.

Traditionally, muscadine grapevines have been trained on overhead trellises; however, they can also be trained on wire trellises like other grapevines, except that yield has been found to be greater with a one-wire type of trellis rather than multiple wires. Muscadine vines are very large, and vines should be spaced 15 to 20 feet apart, with a pollinizing, self-fertile variety for every three female varieties. Vines are pruned more severely than other *Vitis* species. Remove all lateral growth on the main trunk below the wire, and prune back to two to four bud spurs on the remaining growth when the vines are dormant. Muscadine fruit begins to ripen at the end of July and continues to produce through mid-September. The fruit is black or bronze, depending on variety.

Varieties

Muscadine varieties are self-fertile or female. It is important to select at least one of each for fruit production.

Black Beauty Female variety with black skin that is edible. Sweet fruit that ripens in mid- to late season. Very prolific vines. Good cold tolerance and disease resistance.

Carlos Self-fertile variety grown for making jelly and wine rather than fresh eating. Thick-skinned bronze fruit that is bitter.

Darlene Female, bronze-fruited variety that produces large grapes with edible skin. Fruit has a pinkish tinge when fully ripe. Ripens early to midseason. Good disease resistance.

Early Fry A bronze female variety that produces early in the season. Good-quality, large fruit with edible skin. Good cold tolerance and disease resistance.

Ison Black, self-fertile variety that produces large clusters of edible-skinned fruit. Ripens early to midseason. Good disease resistance.

Supreme A black, female variety that produces prolifically on large vines. Skins are edible. Fruit thinning may be required to prevent overcrowding. Ripens mid- to late season. Good disease resistance.

Tara Bronze, self-fertile variety that produces large fruit with good flavor. Ripens early to midseason.

Victoria Red A newly released variety that has shown much promise as being resistant to Pierce's disease and fruit rotting caused by tight clustering. Red grapes grown for fresh eating and blending with other varieties in wine production. Very vigorous vines. Have died from freeze damage in trials in Arkansas and should not be planted north of zone 7b.

JERUSALEM ARTICHOKE
Helianthus tuberosus

❂ **DESIGN USES:** The yellow flowers of the Jerusalem artichoke are a bright addition to the summer garden. Plant tubers in full sun toward the back of the bed where they will not overshad-

Jerusalem artichoke
(Photo by Paul
Fenwick)

ow other plants but will provide a nice, colorful backdrop. Jerusalem artichoke will also make a nice mass planting to be viewed in distant beds.

Description

Jerusalem artichokes are also known as "sunchokes," which may be a more accurate name, as they are not artichokes and originated in central North America. They grow from 5 to 10 feet tall and bloom during the summer months when the rest of the garden may be looking kind of ragged. These delightful members of the sunflower family are grown for their tasty tubers. The texture is crispy when they are raw, or they can be cooked and used in a similar way as potatoes. Jerusalem artichokes are grown from tubers planted in the early spring in well-drained beds and harvested in late fall into midwinter after a few hard freezes. Freezing temperatures improve the flavor. Dig the tubers with a pitchfork, digging under the plants and lifting. Tubers can be a foot under the ground, and if any are left in the ground, they will come back the following spring. Jerusalem artichokes can take over a bed if left unchecked.

STRAWBERRIES
Fragaria spp.

✻ **DESIGN USES:** Strawberry plants can provide a cheerful border for the dreary winter months. The leaves are shiny and green, and bright white flowers appear in the early spring, followed by bright red berries dangling from their stems. For children's gardens, plant large beds of strawberries with pathways for fun harvesting. Be sure to mulch the plants well so the berries do not come in contact with the soil.

Description

Strawberries will be more annual than perennial in the hotter areas of Texas, as they do not like hot weather. Even in cooler areas of Texas, strawberries will wilt and appear stressed. They will appreciate a little shade in the summer when grown as perennials. Otherwise, strawberries do not require a lot of effort to produce an abundant, sweet yield.

There are two types of commonly cultivated strawberry plants, "June-bearing" and "ever-bearing." The ever-bearing varieties fruit later in the season and will produce two or more times per year. The June-bearing varieties are the best choice for hotter areas of Texas. They fruit only one time, producing mostly in the early spring.

Strawberry plants produce runners from the mother plant that will reach out and take root a little distance away. To save the plants for fall planting, clip the runners and plant them in a shadier spot to protect them from the summer heat. With this technique, the plants may survive to be transplanted to a sunny spot the following fall.

Cultivation

Strawberry plants prefer soil that is somewhat acidic. If the existing soil is alkaline, strawberries can be grown in raised beds or in containers, with plenty of well-decomposed compost. Good

Strawberry plant border

148

drainage is very important for strawberry plants. Earthworm castings and rabbit manure compost will help in lowering the pH. Pine straw mulch also helps add some acidity and keeps the berries from coming in contact with the soil. Strawberry plants should always be mulched. This not only protects the berries but keeps the roots of the plants cool and moist. Mulch can also be used to protect strawberry plants from severe winter freezes. Healthy strawberry plants grown in good organic soil do not usually have many insect or disease problems, but mulch protects the leaves and berries from soilborne fungal diseases. Be sure the mulch does not cover the leaves. Keep in mind that the mulch will have to be opened up for rooting runners to take hold. Any plant debris or damaged leaves should be removed to prevent insect and pathogen access.

Strawberry plants are commonly bought as crowns. They will appear as roots with a tiny tuft of growth on top. Strawberries are very appreciative of good organic soils. Pre-prepare the beds with plenty of organic matter and bone meal or other organic form of phosphorus. Strawberries should be planted when temperatures are consistently below 90°F, usually in late September or early October for most areas of Texas. Plants are set out in November in the Gulf Coast region. Strawberry plants can also be planted in early spring; however, they will not be as productive as fall-planted strawberries. Plants should be firmly set at least 12 inches apart in well-drained beds with the tip of the crown above the soil.

Varieties

Allstar Consistently large, orange-red berries produced in late midseason. Good resistance to fungal diseases.

Cardinal Bright red, large fruit produced in midseason. Resistant to airborne fungal diseases.

Chandler A heavy producer with a longer growing season than most June-bearing varieties.

Seascape Ever-bearing but can be a good strawberry for the South because of its low-chilling hour requirements. A good choice for container growing, allowing the plants to be moved to a more shaded location during hot summer months. Good disease resistance.

Sequoia Fruit produced early and is very sweet. Good pest resistance.

Surecrop Medium-sized, bright red fruit produced in midseason. Good resistance to fungal diseases.

Sweet Charlie Very prolific plants that produce very sweet fruit. Produces early in the season and has good disease resistance.

Herbs

✺ **DESIGN USES:** As well as spicing up cooking, herbs season and complete a garden's flavor. The evergreen herbs—bay, rosemary, oregano, marjoram, garden sage, and thyme—can help form the garden's structure. Brightly colored blossoms of Mexican mint marigold, chives, and bee balm bring interest and pollinators to the garden. The interesting textures of fennel, lemongrass, and dill can be used to punctuate garden beds. Thyme, marjoram, oregano, and salad burnet make great border plants. Salad burnet and cilantro are good additions to the winter garden. Variegated thymes and sage add a nice silvery element. Sprinkle herbs liberally through the garden to taste.

ANISE
Pimpinella anisum

This feathery herb in the Umbelliferae family is an annual that can be started from seeds in the spring. Clusters of tiny white flowers grow on stalks in the summer. The lower leaves are rounded with serrated margins. The upper leaves are lacy and fernlike. Anise is grown for its seeds, which are harvested when they turn brown. The seed is used whole or ground and has a licorice flavor. It is used to flavor breads and cakes or to add an interesting zest to applesauce or baked apples.

BASIL
Ocimum basilicum

Basil is a warm-weather annual prized for its unique and pungent aroma and taste. There is no substitute for fresh basil in cooking. Dried will work in the winter, but the gardener waits with

Mint plants (Photo by Andrew Edmonson)

Thai basil flowers (Photo by Andrew Edmonson)

in a sunny window and sct out when the weather warms. Pots of the herb are available in nurseries in the spring and summer.

Sweet basil, probably the most common variety, grows to about 24 to 48 inches in height. It has glossy green, rounded leaves, and the flower spikes have small white blossoms throughout the summer. A purple-leafed variety with ruffled edges on the leaves makes a beautiful accent in the garden. Globe basil is a petite basil that forms a low-growing mound that makes a nice border plant. Lemon basil combines basil and lemon flavors. Thai basil is a common flavoring in many dishes from Thailand. The stems and flowers have a purple color, and the leaf is smaller than the sweet basil leaf. It is very interesting in appearance and flavor.

BAY, SWEET
Laurus nobilis

Bay plants are small evergreen trees that are somewhat winter sensitive in many areas of Texas. They make lovely container plants and are easy to maintain in a pyramidal shape for formal plantings. Container planting is the best way to keep bay trees from being harmed in severe freezing temperatures, especially in the northern parts of Texas. The leaves are soft and bright green when they emerge. They become darker and firmer as they mature. They are somewhat pointed at the tips and have slightly serrated margins. Whole bay leaves are commonly added to soups and stews and removed after cooking.

eager anticipation for the first fresh leaves in the spring. Basil must be protected from freezes. Many gardeners will keep a pot or two during the winter and bring them in when the temperatures dip below freezing. Basil can be grown from seeds, sown directly in the ground, or started indoors

BEE BALM
Monarda didyma

Bee balm is a perennial herb. As the name suggests, this herb attracts bees and butterflies, and it makes a good addition to the garden for increased pollination. Bee balm can be somewhat aggressive. It should be planted in a bed where it will receive plenty of sunlight and have room to spread. The 3- to 6-inch leaves grow on mounds 1 to 3 feet tall every spring. They are topped in the summer with red, pink, or purple starburst flowers. Flower color will vary depending on the selected variety. Bee balm can be grown from seeds; however, it is easier from container-grown plants. The roots can easily be divided to create new plants, which can then be planted in new areas of the garden.

Bee balm (Photo by Jacopo Werther)

This herb has been cultivated in the United States for making tea and seasoning meats since before the first European settlers arrived. Native Americans shared the knowledge of this beautiful herb with the American colonists, who named it "Oswego Tea," for the Oswego Indians. Another common name is bergamot.

BURNET, SALAD
Sanguisorba minor

This low-growing perennial has delicate, rounded leaves with even, scalloped serrations. The cucumber-flavored leaves grow on stems that emerge from a central rosette. It is a mannerly grower and usually stays around 1 foot in height. Salad burnet makes an attractive, low border plant, or it is very pretty on the edge of a container planting where it can spill over the edge. In most winters salad burnet will be evergreen; however, it appreciates some afternoon shade in the heat of the summer. Seeds can be sown directly in the beds in the fall or spring. Container-grown plants are usually available in specialty nurseries. Salad burnet has a wonderful cucumber flavor and is used fresh in salads and herbed butters. This herb makes a nice addition to winter salads when the taste of fresh cucumbers is just a memory.

CHAMOMILE, ROMAN
Chamaemelum nobile

This herb produces a 6- to 12-inch feathery mass of foliage with small, daisylike flowers rising above the foliage on spikes. It can die out in the summer heat but will reseed easily and appear the following year. Seeds are sown directly in the garden in the spring. Chamomile flowers make a fresh apple-scented tea that has been used for soothing upset stomachs and as a sleeping aid for hundreds of years. Harvest the flowers when they are dry, and leave the flowers on a clean surface to dry in the sun.

CHERVIL
Anthriscus cerefolium

Chervil has feathery, bright green foliage that grows about 12 to 18 inches in height. The leaves have a licorice flavor. This is an unusual herb in that it can be grown in part shade. Chervil does better in the cooler months and should be sown in the fall. A long taproot prevents chervil from being grown from transplants. The white umbel flowers should be removed to prolong the season. Chervil leaves are used much like parsley. They lose most of their flavor when cooked, so they should be sprinkled on food when it is almost finished cooking.

CHIVES
Allium schoenoprasum

Chives make pretty border or container plants. Shiny green leaves emerge from bulbs and produce pretty pink-purple, umbel-type flowers in the summer and fall. Their cousin, garlic chives (*A. tuberosum*), has flat leaves and produces white

flowers. Both species are well suited for the Texas garden. Chives can take a little shade but do best in full sun. The plants can be started from seeds but are usually container grown. They can easily be divided in the fall to create additional plants. Chives should be eaten fresh. They are commonly used in egg or potato dishes for a mild garlic flavor and are a nice addition to herb butter.

CORIANDER (CILANTRO)
Coriandrum sativum

Coriander is the same plant as cilantro, the popular leafy herb that gives pico de gallo its unique

Cilantro (coriander) flowers (Photo by Andrew Edmonson)

flavor. It is also called Chinese parsley and is used in many Asian dishes. The seeds are called coriander, and the leaves are called cilantro or Chinese parsley. When in bloom, coriander can grow up to 3 feet in height. The lower leaves have three deep, serrated lobes. The upper leaves are very fine and feathery. Coriander flowers are tiny and white. Coriander should be sown or planted in a well-drained site in the fall. The plants will die in the heat of the summer, but they generally come back from seed the following fall, especially if planted in good garden soil. For coriander seasoning, the seeds should be left on the plant until they turn dry and brown. Then they should be hung upside down in a paper bag until the seed casing splits open. The seeds are then ground into a powder.

DILL
Anethum graveolens

Dill is a great accent in the garden. It has a definite presence, with its 3- to 4-foot-tall feathery,

Dill flowers (Photo by Andrew Edmonson)

blue-green foliage and bright yellow umbel flowers. Dill is a favorite food of many butterfly larvae, especially the black swallowtail (*Papilio polyxenes asterius*), so it is a good idea to plant enough to share. Plant dill near brassica plants, such as cabbage and broccoli, to attract the beneficial wasps that attack cabbage worms. Plant sets in well-drained soil in a sunny location in the spring or fall. Dill can be dried, but the flavor is much more pronounced when fresh. The leaves and flowers are used to give dill pickles their tangy flavor. Diced dill is also a great addition to deviled eggs and fish sauces.

FENNEL
Foeniculum vulgare

Fennel is a perennial with delicate, feathery foliage that is green or bronze tinted depending on the variety. Flat, yellow flower umbels are very dramatic and can be 6 inches across. This plant can reach 5 feet in height and makes a very interesting accent in the garden. Fennel should be planted in well-drained soil in a sunny spot. This plant can deter the growth of other plants, so it should be planted in its own area. Fennel has long been cultivated for the savory anise flavor of its seeds and leaves. It is used to season Italian sausage and is especially good in fish dishes. The bulbs of Florence fennel, or finocchio, are eaten as a vegetable.

Fennel (Photo by Andrew Edmonson)

LEMON BALM
Melissa officinalis

Lemon balm is a perennial with upright stalks that can reach 2 to 3 feet in height. The light green leaves are heart shaped and, not surprisingly, have a wonderful lemon fragrance. In the late summer, lemon balm produces small white flowers that attract bees. The plant will die to the ground in the winter and reappear the following spring. Lemon balm is in the mint family and has similar requirements. This plant can take some shade, especially in the afternoon. It is a durable plant and will thrive in most soils but will do better in a well-prepared bed. Lemon balm does best if started from container-grown plants. This herb imparts a flavor of lemon to teas and fruit salads or can be used as a garnish for summer drinks.

LEMONGRASS
Cymbopogon citratus

Lemongrass is a tender perennial that will probably need to be replanted most years in less temperate areas of Texas. The strong lemon flavor this light green, 2- to 3-foot-tall grass imparts to teas

Lemongrass

and any savory dish where lemon flavor is desired makes it well worth a seasonal investment. The leaves are coarse in texture, especially the outer leaves. When harvesting, select more tender leaves from the inner part of the plant. Plant lemongrass from containers in full sun after all danger of frost has passed. It can also be propagated by dividing the roots if it survives the winter. Prune back the top growth after dividing, and new growth will shoot from the base. Lemongrass contains citronella and would be a good choice around outdoor seating areas to repel mosquitoes.

MARJORAM
Origanum marjorana

Marjoram is a type of oregano. The flavor is a little milder but very similar. It is a somewhat tender perennial that forms a tidy mound of gray-green, oval leaves, reaching about a foot in height. Small white flower clusters form along the stems. Marjoram makes a very nice border or container plant, and can be started from seeds indoors or sown directly in the garden. Marjoram grows best from container-grown "sets." Mature plants can be divided and new plants started by this method. Plant marjoram in a sunny location, although some afternoon shade is acceptable. The soil must be well drained. Marjoram is used in Italian dishes, soups, stews, and other savory dishes in the same way common oregano would be used. The herb keeps its flavor when it is dried, but more will be required than if the fresh herb is used. It is most flavorful at bloom time.

MEXICAN MINT MARIGOLD
Tagetes lucida

This perennial herb forms a low bush in the summer, reaching 2 to 3 feet in height. It has dark green, elliptical leaves that are serrated on the edges. The golden-yellow flowers grow in clusters at the tips of stems. They bloom in the end of summer and into the fall and are smaller and not as showy as those of the better-known garden marigolds. Mexican mint marigold can be somewhat winter sensitive in northern parts of Texas. The plant will die after the first hard freeze and should be cut back to the ground at this time. Plant Mexican mint marigold in well-drained garden beds in an area that receives full sun or a little afternoon shade. Seeds can be started in a sunny window inside or can be sown directly into the soil. Container-grown plants are usually available at local nurseries.

Mexican mint marigold (Photo by Andrew Edmonson)

The leaves and flowers of Mexican mint marigold have an anise flavor and are often used as a tarragon substitute. The flavor is very similar. Leaves are added to vinegars, sauces, and fish dishes. They tend to lose their flavor during cooking, so they should be added at the final stages to cooked dishes.

MINT
Mentha spp.

Mint is a semi-evergreen perennial in areas of Texas that experience freezing temperatures. It becomes somewhat untidy and leggy in the winter and benefits from a good trimming right before the new spring growth. There are many different varieties of mint. Some of the more popular ones include spearmint (*M. spicata*), which grows to 2 feet and has dark green, crinkled leaves. Peppermint (*M.* × *piperita*) is a strongly scented, large-growing mint (up to 3 feet in height) that has small purple flowers. Orange mint (*M. citrata*) grows about 2 feet tall and has a strong orange scent and flavor. It has rounded, dark green leaves growing on red-tinted stems and produces pink to lavender flower spikes. Corsican mint (*M. requienii*) is a small ground cover. Rarely reaching more than 8 inches, it is very pretty planted along pathways or between stepping-stones, especially in locations that receive afternoon shade.

Mint is one of the few edible plants able to take shady conditions, although it will grow more thickly if it has at least morning sun. Mint can also take temporarily wet soil conditions. Mints are

Mint (Photo by Andrew Edmonson)

usually propagated by cuttings or division. They do very well from containers. Mint can be dried or frozen, but the flavors are better fresh. They make lovely garnishes or can be added to desserts or tea.

OREGANO
Origanum vulgare

Oregano is a hardy, evergreen perennial that grows 18 to 24 inches in height. It makes a very nice border or container plant. The leaves are small and oval shaped with a tint of purple. Oregano produces small pink-purple flower whorls on the end of stems. It is best to keep the flowers clipped back to encourage more vegetative growth. Oregano can be planted directly in the garden from containers. Clumps of the herb can also be divided to form new plants. Some people prefer to grow this herb

Oregano

in a pot to keep it contained. Oregano retains its flavor when dried and should be harvested before bloom when the flavor is stronger. The leaves are a primary ingredient in marinara sauces used in Italian cooking. They are also used to season meats, beans, soups, and stews.

PARSLEY
Petroselinum crispum

Parsley is a true biennial that blooms in its second year before it dies. The life of the plant can be extended by pruning off the yellow umbels of tiny flowers. The leaves turn bitter after the plant blooms. There are two varieties of parsley commonly grown, plain leaf (Italian), which has a stronger flavor, and curly parsley. The plain leaf type has flat, fan-shaped leaves, and the curly variety's leaves are ruffled. Both varieties grow to be about a foot tall and make very pretty borders. They can be sown directly in the garden in the fall, but germination is spotty and can take 3 weeks. If the plants are started from seeds, soak the seeds in water for 24 hours before planting to soften the seed coat. Container-grown plants can be set out in the fall or spring and even into the summer. Parsley can take some afternoon shade but prefers full sun. Plant in a well-drained location.

Parsley is used as a garnish or to season many savory dishes. It is a prime ingredient in tabouli. Parsley is also a favorite food of many Lepidoptera species, especially the swallowtail butterfly. This can be a good or bad thing depending on affinity for butterflies and parsley. They are usually fairly easy to remove.

ROSEMARY
Rosmarinus officinalis

Rosemary is a wonderful plant for the landscape. It is one of the few edible evergreen shrubs. The upright variety of rosemary usually grows to about 3 feet, but it can reach 6 feet if it is really happy. There is a prostrate variety (*R. officinalis* 'Prostratus') that grows to only about 2 feet in height but can cascade another 2 to 3 feet over a wall or container. Rosemary has a strong, aromatic pine scent.

Rosemary (Photo by Andrew Edmonson)

Plant rosemary from containers in well-drained soil in full sun. Rosemary cannot tolerate soggy soil conditions and is susceptible to fungal problems from extended periods of high humidity. To keep plants healthy, it is important to prune any dead growth back to the living branches. Sometimes this means pruning all the way to the base. Dead branches are commonly covered by living branches. Lift top growth to inspect for dead limbs. It is also important to be careful not to cover branches with mulch. The branches will often touch the ground, and it is easy to accidentally cover them.

Because rosemary is evergreen, it may be harvested year-round. Rosemary has a strong flavor, and a little goes a long way. Sprigs are commonly burned on a grill fire to season meat. It is also a very useful flavoring for fowl and lamb dishes or to make rosemary potatoes.

SAGE
Salvia officinalis

Garden sage is an evergreen perennial that grows to about 2 feet. It has soft, gray-green, elliptical leaves and produces small, light blue to purplish flower whorls at the ends of stalks. The flowers are good for attracting pollinators. A tricolor variegated variety has cream and pink mixed in with the green. Pineapple sage (*S. elegans*) is a tender

Sage plant with praying mantis (Photo by Andrew Edmonson)

perennial in much of Texas and will die if temperatures fall to 20°F. For this reason it is an annual in many areas of Texas. This herb has bright red, tubular flowers that attract bees and butterflies. The leaves have a strong pineapple scent and are used to garnish fruit drinks, or they can be chopped and added to fruit salads.

Sage should be planted in full sun in well-drained soil. It can be started from seeds, but it takes about 2 to 3 weeks for germination, and there is a danger of the seedlings dying from a fungal disease called "damping off." It is safer to set out container-grown plants. Sage is not a long-lived perennial and will probably need to be replaced every few years. Garden sage is very pungent and is the primary flavor in stuffing or dressing. It can also be used to flavor fondue.

Variegated and garden thyme

THYME
Thymus vulgaris

Thyme is a low-growing, evergreen perennial, perfect for border plantings or spilling over the rim of a pot. Garden thyme has small gray-green leaves and produces purple-pink flower whorls at the ends of the stems. Bees are very attracted to thyme, making this the perfect herb to plant as a border to attract pollinators. The 'Elfin' variety has very tiny leaves and grows only a few inches in height. Lemon thyme has a mild lemon scent and flavor. Its leaves are variegated with yellow margins for a very pretty contrast against darker-colored plants. Plant thyme in a well-drained bed in full sun. The plants can be grown from seeds; however, they will be more reliable if they are set out from containers. Harvest the leaves before the plants flower for the best flavor. The leaves are used to flavor soups and stews and are a wonderful seasoning for meats and steamed vegetables.

✸ Vegetables

The vegetables included in this section are annuals, and because the members of a family will generally share the same cultivation requirements, they are grouped by family. It is best to avoid planting the same plant or plants in the same family in the same spot for 2 to 3 years. There is a greater possibility of soilborne diseases and insects specific to a family harboring in the bed from year to year. Alternating locations is a method of outfoxing these pests.

Chinese cabbage in flower

There are too many varieties available to provide a comprehensive list. Varieties listed in this book are selected for unusual physical characteristics, such as leaf and fruit color, as well as their adaptability to growing conditions in Texas. The reader is encouraged to do further research on the many varieties available through seed catalogues and local nurseries.

There are benefits in selecting heirloom seeds. They may have special characteristics in flavor or appearance that are not available in more contemporary varieties. Many of the more commonly grown seeds are hybridized, which prevents a gardener from saving the seeds from year to year. These hybridized varieties do not "come true" from seed and cannot be relied upon to produce the same quality of fruit from year to year. Open-pollinated and heirloom varieties reproduce similar progeny reliably and are the varieties selected for this book.

Saving seeds can be very rewarding for many gardeners, and the seeds should become more improved for the site conditions from year to year, as seeds are selected from plants with the most desirable genetic traits. Ideally, these seeds will be more disease and insect resistant through genetic selection. More detailed information on seed saving, as well as additional hybrid varieties, can be found in the appendices.

Beet Family

Description
Amaranth, beets, spinach, and Swiss chard are durable, nutritious, and beautiful. Gardeners will be

Swiss chard (Photo by Andrew Edmonson)

delighted with the cheerful, bright colors, used as accents or mass plantings. These plants, except the tall stalks of amaranth, also make attractive container plants. Swiss chard and beets are cultivars of the same plant, grown for different purposes. Over time, beet plants have been selected for their edible roots as well as their foliage. Swiss chard does not develop a fleshy root and is grown for its edible leaves. Swiss chard will usually survive freezing temperatures and can continue to provide bright, colorful accents for Texas gardens year-round in some areas of Texas.

Cultivation
Chard and spinach will perform better than beets if started in containers. Beets require plenty of room for the roots to develop and will do best if sown directly in the bed. Plant seeds in full sun ½ to ¾ inch deep after average daytime temperatures reach 80°F in the early fall. Sometimes spinach

will work as a spring planting; however, it is more likely in Texas that warm spring weather will cause the plants to bolt and become bitter before the leaves can be harvested.

Seeds of these beet family plants are produced in dried seedpods called achenes along flower stalks when the weather warms. These achenes are a fruit that may contain a number of seeds in the pod. The seedpods remain intact at maturity, so they will produce a number of plants when they are sown. Seedlings will need to be thinned after they emerge. The simplest way to do this is to clip the additional seedlings with scissors rather than try to pull them by hand, which will disturb the delicate roots of the seedlings.

AMARANTH
Amaranthus spp.

Amaranth growing in residential landscape, Austin, Texas

❋ **DESIGN USES:** Amaranth is often grown as an ornamental. The brightly colored seed heads, as well as the dramatic foliage color of some varieties, make this plant a real showpiece of any edible landscape. Because they are tall, it is best to place amaranth plants toward the back of the garden. Use as accents or en masse.

Description

Amaranth is a beautiful and nutritious grain that is finding resurgence in the edible landscape. Amaranth is a traditional grain that has been cultivated in the Americas for thousands of years. There is evidence that amaranth was a major food of the Aztecs, as well as other Native Americans. This grain contains 14 to 16 percent protein and is high in lysine, an amino acid commonly found in legumes but not generally in grains. Drought tolerance and adaptability make this plant a good selection for Texas gardeners who would like to add a healthful, low-maintenance grain to their garden.

Amaranth seed heads can be quite dramatic and colorful, creating beautiful flowerscapes in various shades of green, maroon, pink, and gold. The plumelike foliage grows on tall stalks that can reach from 2 to 8 feet. Seed heads vary in size from 4 to 12 inches long and 2 to 8 inches wide. Amaranth is grown commercially for seed in the United States. The leaves are also edible and are a valuable leafy green for the summer when most other greens are no longer producing. People with kidney problems should take care not to overdo consumption of the leaves, because amaranth, like spinach and many other leafy greens, is high in oxalic acid.

There are more than fifty species of amaranth in the United States. Some are grown for their ornamental seed heads, such as the ornamental cockscomb. *A. tricolor* and *A. dubius* are Asian varieties grown for their edible leaves, and *A. cruentus* is grown for both the leaves and the grain. The most commonly cultivated species in the United States is *A. hypochondriacus*, grown primarily for the seed heads. Certain wild varieties of amaranth are considered to be weeds. *A. palmeri* (Palmer's pigweed) is an invasive weed in crops because of its resistance to herbicide application. Pigweed is also toxic to grazing animals, so ranchers are not too fond of this plant either. For gardeners, amaranth can be a beautiful and healthy addition to the garden.

Cultivation

Amaranth is a warm-weather crop that is seeded directly in the soil in the spring after any danger of frost. This plant will respond well to additional nitrogen in the form of well-decomposed compost worked into the top few inches of the soil before planting; however, do not overdo it. Heavy nitrogen application can cause the plants to "lodge," or become top-heavy and fall over. It is best to plant the seeds about ½ inch deep and cover loosely. The tiny seeds will struggle to emerge if the soil is packed too tightly. Although it grows fairly slowly at first, after it reaches 6 to 10 inches, amaranth literally grows "like a weed." Be sure to give the individual plants room to grow. Overcrowded plants can become leggy and produce spindly seed heads.

Amaranth is generally disease-free if planted in well-drained soils. Fungal diseases can attack these plants, especially at the soil level in poorly drained soils. If this happens after plants have matured, it can cause the tall stalks to lodge.

Amaranth seeds will mature at different rates. The early seed heads can be harvested by gently rubbing them over a tarp or paper bag. Late seed heads can be left to dry on the stalks and harvested before they shatter. It is best to harvest the seeds after a killing frost if possible. This will kill off any harboring insects that might be a problem during storage. Be sure that seeds are completely dry before storing.

Varieties

Elephant Head A 3- to 5-foot plant with an unusual red seed head that appears to have a long "trunk" like an elephant's.

Golden Giant Produces large, prolific, golden seed heads on 6-foot stalks.

Greek Giant Reddish-pink seed heads on less-branched stalks 3½ feet tall.

Green Callaloo Grown for its leaves, which taste similar to a spicy spinach and commonly used to make Caribbean callaloo seafood soup.

Hartmann's Giant A striking plant with bright red seed heads on top of 10-foot stalks.

Hubei Grows to about 4 feet in height and has red seed heads.

Love-Lies-Bleeding Produces long, ropelike, exotic-looking seed heads. Green and red varieties. Grown for seeds and leaves.

Mayo Indian Red stalks and seeds with green leaves that have red veins. Grown for both leaves and seeds.

Miriah Grown for greens. Golden-brown seed heads on 5- to 6-foot stalks.

Molten Fire A striking variety, 4 feet tall, with bright red leaves on the top, turning to burgundy-red leaves on the lower part of the plant. Grown for the edible leaves.

Orange Giant Orange-yellow seed heads on 6- to 8-foot stalks.

Polish Burgundy-red leaves and seed heads on 3- to 4-foot stalks.

Thai RW Tender Grown for its edible green leaves. Popular in Thai cooking.

Vietnamese Red Dramatic, 6-foot plants with reddish-green leaves and red flower heads.

BEET
Beta vulgaris

❊ **DESIGN USES:** Even though beets have red leaf veins, they are not as showy as their Swiss chard cousin. They can make a nice low planting with taller plants punctuating the bed. The gray-green foliage of artichokes and cabbages are nice choices for this purpose.

Description
Beet roots come in a variety of colors, including orange, red, purple, yellow, white, and even a

Beet plants

striped red-and-white variety called 'Chiogga.' They can also be round, cylindrical, or tapered like a carrot. It is a shame that these beautiful roots have to grow under the ground. The part of the plant that grows above the ground is a leafy green that is generally green with a reddish tinge. Beet tops are also edible. The leaves can be harvested as the plant matures, taking the lower leaves first and allowing the other leaves to mature. When storing beets, it is important to leave 1 or 2 inches of the leaf stem on the beet root to prevent them from "bleeding" their bright red juices.

Cultivation
Beets need to be planted in deep, rich garden beds to allow plenty of room for the roots to develop. A

loose sandy loam with a pH between 6.2 and 6.8 is ideal. East Texas gardeners may need to add lime to the soil to reduce pH. Beets should be thinned 3 to 6 inches apart, and small or "baby" beets can be harvested to allow more room for full-grown beets to mature. Beet roots can reach depths of 18 to 24 inches, and they are heavy feeders. Prepare beds with plenty of well-decomposed compost before planting. Beets can suffer boron deficiency, causing them to turn brown and have hollow cores. A soil test should reveal this deficiency before planting.

Varieties

Albino Pure white, round, and sweet.

Bull's Blood Deep burgundy-red leaves. Roots best harvested when young.

Chiogga Round roots with concentric rings of white and red. Interesting on the table, with a sweet flavor.

Crapaudine Long, cylindrical with rough, dark skins and dark red, very flavorful flesh.

Cylindra Dark red beets 8 inches long and 1½ inches wide. Very good flavor.

Detroit Dark Red Dark red with roots 2½ to 3 inches long. Stores well. Resistant to downy mildew.

Flat of Egypt Fast growing with flattened, 3-inch crimson-purple roots and short tops.

Golden Golden colored and fast growing. Resists becoming fibrous.

Lutz Retains sweetness and texture even though fairly large. Stores well.

SPINACH
Spinacia oleracea

✿ **DESIGN USES:** Glossy green spinach leaves are a cheerful backdrop for brightly colored foliage, flowers, and fruit. Use as a border or planted in swales. Plant full beds of spinach punctuated with Swiss chard, Chinese cabbage, or bold-textured kale, and add pockets of red-leafed lettuce or mustard greens.

Description

Spinach can be grown in the spring or fall in Texas; however, spring plantings may be a little more difficult. Spring weather often alternates between periods of warming and cooling that can confuse the plants into thinking the seasons have changed. This causes them to bolt and go to seed prematurely.

Cultivation

For fall planting, keep in mind that spinach does not germinate well until soil temperatures are below 70°F. Soil temperatures can be cooled with the use of shade cloth, and the seedlings will appreciate a little extra shade until daytime temperatures cool. Soils with a neutral to high pH are preferable for spinach growing. Luckily, much of Texas can provide these conditions. East Texas soils may need additional lime to reduce pH. Before planting, it is helpful to soak the seeds overnight to start the germination process. Seeds can be sown in full sun or part shade in successive plantings every 10 to 14 days to extend the season. Spinach likes rich, well-drained soil and sufficient moisture to keep the plants moist but not wet. Weekly applications of fish emulsion and seaweed will be rewarded with healthy, productive plants. The season can be extended through the winter if the plants are covered by a loose mulch, such as straw, during hard freezes and uncovered when the weather warms. Spinach will survive temperatures to the mid- to low 20s.

Varieties

Spinach cultivars fall into three types, based on how crinkled, or savoyed, the leaves are: smooth, savoyed, and semi-savoyed. Spinach types will cross-pollinate readily.

Young spinach plants

Bloomsdale Crinkled, glossy green leaves more resistant to heat and cold than leaves of some varieties.

Bloomsdale Long-Standing Glossy, heavily savoyed leaves with an upright habit. Good heat and cold tolerance.

Giant Noble Very large leaves (up to 25 inches) that remain tender and flavorful.

Giant Winter Medium-green, semi-savoyed leaves with good flavor and cold tolerance.

Malabar *Basella alba*, not a true spinach. Fast-growing tropical vine that is a good leafy green for summer. Red stems and leaf veins add an interesting vertical element for trellises. Good plant to follow spring pea plantings.

Merlo Nero Dark green, savoyed, Italian variety.

Oriental Giant Very fast growing with heavy yields. Grows 12 to 15 inches with smooth, arrow-shaped leaves. Some mildew resistance.

Springer Upright, fast growing, semi-savoyed, deep green, shiny, arrow-shaped leaves. Some disease resistance. Slow to bolt.

Viroflay Giant Can reach 2 feet in diameter. Thick leaves up to 10 inches long and not bitter.

Winter Bloomsdale Dark green, heavily savoyed leaves. Resistant to blue mold, blight, and mosaic virus. Slow to bolt.

SWISS CHARD
Beta vulgaris var. *ciclis*

❋ **DESIGN USES:** The vivid colors of the stalks and midribs of Swiss chard, as well as the bold texture of the foliage, make this plant a great addition to the designer's palette for an edible garden. Use as a border plant, as an accent plant, or in little pockets to add color and interest to the edible landscape. Swiss chard also makes a nice container plant.

Description
Swiss chard is a truly wonderful performer for Texas gardens. The 18- to 24-inch-tall plants form accents of glossy leaves that are bold and beautiful, as well as highly nutritious. The stalks of the 'Rainbow' variety have a very striking display of

Swiss chard (Photo by Andrew Edmonson)

bright red, yellow, gold, orange, pink, and violet color. There are also green-stalked varieties. They all perform well and add interesting color and texture to planting beds.

Cultivation
Swiss chard grows best in the cooler months in Texas. Although growth will be slowed in the hotter months, chard can continue to survive even during summer weather. Chard can be seeded directly in the garden or set out as transplants when temperatures begin to cool. Spring sowings should begin 3 to 4 weeks before the last average frost.

Swiss chard plants generally survive freezing temperatures and continue to provide nutritious greens all winter. The seedlings should be thinned to 1 to 1½ feet apart in well-worked garden beds. Chard will appreciate additional compost worked into the soil. The leaves can be harvested as needed, taking the outer leaves and allowing the inner leaves to grow.

Varieties
Flamingo Pink Bright pink stalks that stand out in the garden.
Fordhook Giant Ruffled, green leaves on thick, white stems.
French Heavy yields of tender, green stalks.
Golden Sunrise Golden stalks with green, crinkled leaves. Leaves hold their color well when cooked.
Lucullus White stalks and light green, heavily crinkled leaves.
Oriole Orange Orange stalks.

Rainbow Provides a striking display of colorful red, yellow, pink, orange, and white stalks.

Ruby Red Bright red stalks with leaves that are dark green and crinkled, like spinach leaves.

Schnittmangold Gelb From Switzerland. Large, smooth, green leaves on white stalks.

CARROT
Daucus carota

☀ **DESIGN USES:** Carrot plants have fernlike foliage that makes a nice contrast with bold-textured plants. Plant en masse in well-prepared beds for a feathery-textured ground cover.

Description

Carrots are fairly easily grown in Texas gardens, and the bright green foliage adds a lacy texture to the garden. The orange vegetable is a taproot and is not visible until the carrots are pulled during harvest. The delicate leaf texture can be a pretty contrast when used as low planting in the front border or as a ground cover with plants of contrasting texture. The most important thing in growing carrots successfully is to be sure the planting bed is loose and well drained to the mature root depth. Even small stones can cause the carrot roots to fork or become misshapen, and heavy soils will hinder the long taproots from forming.

Even though they are usually grown as annuals, carrots are actually biennials that produce flowers and seeds during their second year. Erratic weather temperatures in Texas can often fool

Carrot plants (Photo by Andrew Edmonson)

the plants into thinking they have actually gone through two growing seasons, causing them to flower in the first year.

Cultivation

By providing deep, loose soil and even watering, Texas gardeners can be rewarded with an abundance of bright orange carrots from the fall and winter garden. Carrots grow best in a soil with a pH between 6.5 and 7.5. East Texas gardeners may need to lime their soil during bed preparation. Beds should be loose and free of rocks or other obstacles that could prevent root growth. Compost will help loosen the soil; however, too much nitrogen in the soil will cause the leaves to grow rather than the roots. Carrots grow best in cool temperatures between 40°F and 85°F and will mature in about 65 to 90 days, depending on variety. Carrots cannot withstand temperatures below 20°F. The plants will appreciate a loose mulch when they are 4 to 6 inches tall to protect the roots from freezing temperatures and keep the soil moist.

Carrot seeds are very tiny and are sown directly into well-prepared beds. The seeds should be firmed onto moist soil and covered lightly with screened compost or topsoil. Water the seeds in at planting time with a gentle spray of water. Germination occurs in 1 to 3 weeks, depending on soil temperature. It is important that the seeds not dry out during the germination process. It is difficult to plant the tiny seeds far enough apart, so they will need to be thinned to a few inches apart after they have emerged. Larger varieties will need more "shoulder room" than smaller varieties. Soil moisture should be maintained evenly through the growing season, especially during the first 4 to 6 weeks after emergence. Carrots can become tough and flavorless if they dry out. It is also important that the soil drains well and water does not stand around the roots.

Varieties

There are long and short varieties of carrots, and the roots come in an array of colors, from the common orange to shades of white to a dramatic purple. Carrots will cross-pollinate with other carrot varieties and with the wild carrot Queen Anne's lace. The perimeter of the garden should be kept free of weeds, and different varieties should be kept in different areas of the garden to prevent cross-pollination. This will help ensure that succeeding plantings will be from pure seed.

Belgian White A large, white carrot, 8 to 10 inches long. Almost coreless with a mild flavor.

Chantenay Red Core A blunt carrot about 5½ inches long and 2½ inches wide with a deep orange core.

Cosmic Purple Purple on the outside and orange and yellow inside, growing to 7 inches. Flavor somewhat spicier than that of other carrots.

Danvers 126 Dark orange, 6 to 7 inches long, with smooth skin. About 2 inches wide at the shoulder, tapering to a blunt point. More heat tolerant than most.

Danvers Half Long Deep reddish-orange roots 6 to 8 inches long.

Healthmaster A very late variety, taking 130 to 150 days to mature. Very large, up to 10 inches long, with good flavor and color. Stores well and resists cracking.

Imperator Orange-red roots 7 to 9 inches long. Sweet, tender, and coreless.

Lunar White A white, 8-inch, mild-flavored carrot.

Nantes Orange root 7 inches long.

Nantes Half Long A slender, 7-inch carrot that is nearly coreless.

Oxheart Orange carrot up to 1 pound, 4 to 5 inches long, and 3 to 4 inches wide.

Parmex Small, round carrots sold in French markets. Tops grow to about 10 inches. Diameter usually 1 to 1½ inches. Good crack resistance.

Purple Dragon Reddish-purple, 6-inch carrot with yellow-orange core.

Royal Chantenay Short and blunt in form.

Thumbelina Early small, round carrot with smooth, bright orange skin. Good "baby carrots" when harvested early.

White Satin White, thin, tapered carrot up to 8 inches long.

Yellowstone Yellow, mild-flavored, 8-inch carrot with good texture.

CORN
Zea mays

❋ **DESIGN USES:** Corn produces large, straplike leaves that grow from a center stalk. These stalks can be very dramatic in an edible landscape and can be structural as well, providing support for vining plants. A symbiotic planting of corn, beans, and squash is a traditional planting of indigenous American Indians called "The Three Sisters." The corn is used as a support for the vining beans, which fix nitrogen in the soil, and the squash is used as a "living mulch" ground cover to cool and shade the soil. Plant corn in blocks that are at least four rows wide. The wider the blocks are, the more yield will be achieved. Corn can be planted in blocks to create fun mazes if there is sufficient space in the garden.

Description
Like other edible members of the Poaceae family, corn requires a lot of room to produce sufficient yield. The stalks generally produce one ear per stalk, although some varieties will produce two ears, or a second crop of "baby corn." This small yield per area planted is a consideration if garden space is limited. The Poaceae family includes many of the most commonly grown edible plants. Corn, wheat, sorghum, oats, and rice are all members of this family, but corn is the most common for small acreage. Because of its many uses, from food to silage crop to biofuel, corn is the most widely cultivated of all Poaceae plants in the United States.

As popular as corn is in the American diet, few people have experienced the flavor of freshly picked ears. Corn is at its sweetest within a few hours of harvest, as sugars in the corn do not take long to convert to starch. Freshly picked corn is a real treat and is well worth the effort if space allows.

There are different types of corn, which are grown for different purposes. Sweet corn, as the name implies, is grown for its sweet, tender kernels. Dent corn, also called field corn, is grown for animal feed or for processing into cornmeal or other foods. Dent corn has a tough endocarp around the tender kernel when it is fully mature. This tough shell is removed through liming or by grinding dried corn. Dent corn can be harvested at an immature stage and used for "roasting ears." There are also novelty corn types, such as popcorn or ornamental corn. Most

Cornstalks

of the sweet corn varieties grown in America are from hybrid seeds, developed primarily for sweetness. There are some open-pollinated varieties from which the seeds can be saved for planting the following year.

Cultivation

Corn is planted in blocks of at least four rows to encourage pollination. The pollen is produced in the top tassels, and the wind blows it on to the silks of the ears below. The pollen-receptive silks are the stigmas (or styles) that carry the pollen down to the ovules. Each silk leads to an ovule that becomes an individual kernel in the ear. Because the pollen-producing tassels mature at different rates than the silks, pollination is increased if the stalks are planted in blocks. It is common that stalks on the outer edge of a block will not have as high a pollination rate as the interior plants. Corn will cross-pollinate between varieties, and it is common that kernels will be a mixture of genetic parentage if multiple varieties are planted in the same area.

Plant corn from seeds in the late spring after all danger of frost has passed. The seeds will not germinate if the soil temperature is too low or if the soil is too wet. An additional planting in early to mid-August may yield a fall corn crop if temperatures cooperate. Sugar level is reduced by high temperatures. Corn seeds can be inoculated with a foliar application of a fungus called T-22 (*Trichoderma harzianum*) to help suppress harmful fungi, such as *Pythium, Rhizoctonia, Botrytis*, and *Fusarium* species. Crop rotation and proper fertility management will help prevent most diseases. Corn grows best in soils with a pH between 6.0 and 7.0. Corn is a heavy feeder, and the seeds should be planted in hills or raised beds that have been prepared with an abundance of well-decomposed compost to a depth of 6 to 8 inches.

Sow seeds 3 to 4 inches apart when all danger of frost has passed. Young plants should be thinned to 8 to 12 inches apart after they emerge. It is a good idea to side-dress the plant with additional compost about the time the sixth leaf emerges. The plants will begin to feed even more heavily at this time. Additional sources of nitrogen commonly used by organic farmers include cottonseed meal, feather meal, and fish meal. These will be good amendments if there is a plentiful source nearby. Otherwise, compost will do. Corn plants will also appreciate additional applications of phosphorus. Blood and bone meals can be added to the planting bed at the time the seeds are planted to supply additional phosphorus and nitrogen. Colloidal phosphate is often used as a source of phosphorus in commercial operations, because of the higher cost of blood and bone meal; however, the blood and bone meal have higher levels of available phosphorus, and these products are well suited for home gardens. Colloidal phosphate may be a better choice in areas with heavy clay soils, as this product contains calcium, which helps break up heavier soils. To produce full ears, it is important to make sure the plants receive sufficient, even watering, especially during and after pollination. Lack of water can prevent the ears from filling out. An application of mulch will help conserve water, keep the soil evenly moist, and remove weeds.

Varieties

Some corn varieties can be sensitive to day length and will not produce flowers if the day length is greater than 13 hours. This occurs during the third week of April in Central Texas, so early varieties will be more likely to set silk before the days become too long in most areas of Texas.

Corn is a very important crop. Some might say it is the most important crop in the world. Formidable efforts have been invested in the hybridization of this plant to produce sweet, plump corn for fresh eating. In keeping with the emphasis on open-pollinated varieties that this book promotes, hybrid varieties have not been included. A few sweet corn, open-pollinated varieties have been included, but it is important to note that these suggested varieties are limited for those who may want to explore hybrids for enhanced qualities.

Ambrosia A midseason, sugar-enhanced variety. Stalks grow 6 to 7 feet tall and produce 8-inch bicolor ears.

Ashworth An early-maturing variety, requiring only 69 days to harvest. Can be planted in cooler soils as soon as all danger of frost is passed. The 5-foot stalks produce one or two ears each that are 6 to 7 inches long.

Baby Corn An open-pollinated variety harvested early for 5-inch, entirely edible ears. Ready for harvest in 65 days, 5 days after the silks appear.

Buhl Produces one to two ears each of sweet yellow corn on 6- to 7-foot stalks.

Country Gentleman A white corn variety that grows 6 feet tall and produces 8-inch, tightly packed ears with irregularly patterned kernels. Popular for creamed corn recipes. Requires 90 days to mature so may not have time to develop before the 13-hour day-length window.

Golden Bantam Probably the most commonly grown open-pollinated sweet corn variety. Sweet, plump kernels on 7-inch ears. Eat soon after harvest to preserve sweetness.

Tight ears provide some resistance to borers. Matures in 80 days.

Painted Mountain Multicolored 6- to 7-inch ears that make beautiful fall decorations or can be ground for cornmeal. Plants average 4 feet tall and mature in 85 days.

Rainbow Indian A flint corn variety that produces colorful kernels used for cornmeal. Matures in approximately 105 days, so silks may not mature before the 13-hour day-length window.

Strawberry A variety grown for its 2-inch ornamental ears. Can be cooked like popcorn when dried.

Texas Honeyjune A white corn that matures in 97 days with very sweet flavor. Very tight husks that help prevent earworm infestations. Stalks 8 feet tall and can produce one to three ears each.

Trucker's Favorite A midseason dent corn that produces 8- to 12-inch ears with white kernels. Recommended for roasting or can be harvested early for fresh eating.

True Platinum A white sweet corn variety that grows on 6-foot-tall, purple-tinged stalks. Ears 7 to 8 inches long, commonly two per stalk. Some resistance to blight, rust, and drought. Should be eaten soon after harvest for optimum sweetness. Matures in 78 to 84 days.

Wampum Produces 5-inch ears with varied bright colors. Some husks are colored.

CRUCIFERS

Description

Members of the Brassicaceae or mustard family are also known as crucifers. Leaves of these plants are generally gray-green with white or reddish ribs and veins. If allowed to "go to seed," plants in this family produce stalks with masses of tiny yellow or white flowers. The Brassicaceae family includes many wild plants in Texas, as well as the more well-known cultivated types: broccoli, brussels sprouts, cabbage, cauliflower, greens, kohlrabi, radishes, and turnips. Brassicas cross-pollinate freely with both wild and cultivated varieties. It is important to keep this in mind when desiring to maintain seed purity for future planting.

Chinese cabbage in flower

Cultivation

Brassicas perform best when planted in the fall in Texas. They can also be planted in late winter or very early spring for spring harvest; however, the danger that freezing temperatures will harm tender seedlings and that warm spring temperatures will cause plants to go to seed are good arguments for fall planting. It is important to rotate the location of brassica plantings with other plant families to prevent repeated attacks from pathogens that may still be lurking in the area, even after the plants have been removed. Some brassicas, such as brussels sprouts, broccoli, and cauliflower, are commonly planted from nursery-grown containers. Some, such as kohlrabi and radishes, are more commonly sown directly in the garden. It is important to buy certified disease-resistant seeds whenever possible. Seeds will germinate in about a week under optimal conditions, when soil temperature is around 75°F. When the first true leaves appear, water around the plant's roots with fish emulsion and seaweed and reapply every 3 weeks.

Cruciferous plants prefer a soil pH between 6.0 and 7.5. East Texas gardeners may need additional lime application to raise the soil pH. Brassicas are heavy feeders, requiring ample supplies of nitrogen for good yield. Prepare beds with well-decomposed compost worked into the top 6 to 8 inches of existing soil before planting to provide additional nitrogen. If container-grown plants are set out, compost can be worked into the beds at the time of planting. If direct seeding, work compost into the soil a week or two before planting.

Well-mulched plants will generally survive freezing temperatures. In extreme cold the plants may freeze, destroying the cell walls and structure of the plants. Brassicas are susceptible to a number of fungal diseases, so be careful not to place mulch on the leaves or next to the stems. Floating row covers are useful to protect plants if temperatures drop into the 20s. These lightweight fabric covers are designed to "float" above the plants, or they can be supported by arched frames anchored in the ground. Brassicas can be protected from moth and caterpillar damage with row covers as well. Row covers can also prevent moths from laying their eggs on the leaves, thus preventing their voracious caterpillar progeny from stripping the stalks in a matter of hours. It is best to have the supports in place before freezing temperatures or insect infestations are threatening.

BROCCOLI
Brassica oleracea

❊ **DESIGN USES:** Broccoli plants are pretty in small groupings in the garden. The gray-green foliage makes a nice combination with red- and-purple foliaged plants.

Description

Broccoli does best when container-grown plants are set in the beds rather than direct seeding. Starting seeds indoors gives them more time to mature before fluctuations in the weather cause the plants to bolt in the spring or become stunted from freezing weather if planted in the fall. The key to lush heads of broccoli is to grow a strong base plant.

Cultivation

Prepare beds with plenty of well-decomposed organic matter and fertilize with fish emulsion and seaweed. Compost tea will increase soil microbial activity. These efforts, along with consistent watering, will help develop strong plants with fuller heads.

Check the number of days to maturity. Fall plantings are generally more successful than spring plantings. Water the beds, making sure the soil is not soggy, and then space plants 6 to 8 inches apart. Set transplants deeply, up to the first true set of leaves to give them added stability. Start the transplants off right with a soaking of fish emulsion and seaweed.

Broccoli is shallow rooted and requires consistent watering throughout the growing season. A 2- to 3-inch layer of mulch will help conserve soil moisture and keep the roots cool in warmer temperatures and protected in cold temperatures. To protect broccoli from severe winter temperatures, pile a layer of loose mulch around the plant and pull it back as weather warms. Be careful not to leave mulch directly on the leaves or stem.

After the main head is harvested, smaller side shoots will usually be produced. Do not delay in harvesting. When the heads are mature, the buds should still be tightly closed. Broccoli is most flavorful when the buds began to swell but before they open up to produce seeds, when they may turn bitter. Even though the head is most commonly eaten, the leaves and stems are edible as well.

Broccoli (Photo by Andrew Edmonson)

Varieties

Broccoli Raab Also referred to as rapa, rapine, rappone, fall and spring raab, or turnip broccoli. Raised for its greens rather than its small heads. Commonly used in Italian cooking.

Calabrese (Italian Green Sprouting) A blue-green variety that is easy to grow and has a very good flavor. Grows well from seed sown in the garden.

De Cicco Bluish-green variety with a mild flavor and tender stems. Sends out side shoots for additional harvest.

Purple Sprouting Purple-headed variety. Very cold tolerant and needs to be planted in the fall to overwinter. Large heads and multiple side shoots that form in early spring. Purple stems and heads turn green when cooked.

Romanesco Resembles a light green whorled cauliflower and is a sure conversation piece at the dinner table. Whorls form a cone shape, with tighter and tighter circles toward the center of the head, similar to a nautilus shell.

Waltham 29 A low, compact, slate-green variety. Especially good for fall planting. Somewhat drought tolerant. The 4- to 8-inch heads slow to go to seed. Produces lateral side shoots 6 to 8 weeks after main head harvested.

BRUSSELS SPROUTS
Brassica oleracea var. *gemmifera*

✽ **DESIGN USES:** Brussels sprouts plants can be a dramatic and colorful addition to the garden.

Brussels sprouts (Photo by Andrew Edmonson)

The bold-textured plants are striking when used as punctuation in a sea of colorful lower planting. The purple foliage of 'Falstaff' is very pretty with contrasting yellows.

Description
Brussels sprouts produce bold, oval leaves with colors ranging from gray-green to gray, dark green, and purple, along 2- to 3-foot stalks. The sprouts form in the leaf axils along the stalks and resemble small cabbages.

Cultivation
Brussels sprouts require a long, cool growing season, and planting container-grown transplants in late summer or early fall is the only option for Texas gardeners. If the sprouts form during hot weather, they will be loose and have a bitter flavor. Six-week-old transplants with six to eight true leaves should be planted in the garden 8 to 10 weeks before the estimated freeze date. Set the transplants about 18 inches apart and water with fish emulsion and seaweed.

Brussels sprouts grow best in pre-moistened, deeply prepared soils. Prepare soils prior to the planting date with plenty of well-decomposed compost and additional soil amendments to adjust soil fertility and pH, as needed. Brussels sprouts require consistent water and cool temperatures to produce well. Uneven watering will prevent the plant from producing quality sprouts. Spread 2 to 3 inches of mulch around the plants to conserve water and protect the roots. Transplants may need to be shaded until daytime highs fall below 85°F.

This can be accomplished by planting in the shade of taller plants that will be spent by the time the brussels sprouts mature or by making shade structures over the tender seedlings. Just be careful to provide good ventilation.

The sprouts will mature in 90 to 100 days and should be harvested when they are still firm, taking the lower sprouts first. Head formation can be encouraged by removing the leaves as the sprouts form. It is important to leave a few leaves for plant photosynthesis. Some gardeners remove the top growth of the plant at the final stages of harvest to send all of the plant's energy to the sprouts.

Varieties

Catskill Semi-dwarf, 20- to 24-inch plants. Dark green heads 1¾ inches in diameter.

Falstaff About 24 inches in height with a beautiful purplish cast to the leaves. Striking rich purple heads 1½ inches in diameter.

Long Island Improved Very productive variety that matures over several weeks. Sprouts 1 to 1½ inches in diameter that freeze well.

Red Rubine Deep purple to reddish sprouts with a deep, rich flavor.

CABBAGE
Brassica oleracea var. *capitata*

❋ **DESIGN USES:** Cabbages are a great accent in the garden. Their low growth habit and bold texture make them very pretty nestled among other plants of contrasting color and form. Some varieties have ruffled foliage or variegated veins that add

Red cabbage

another dimension to this beautiful plant. Cabbages can be grown as container plants.

Description

There are two types of cabbages commonly grown in Texas, the "head" types and the "Chinese" varieties (*Brassica rapa*). Head cabbages form balls of tightly closed leaves with the outer leaves forming an open nest. These types of cabbages are either green, purple, or savoy. The savoy type has ruffled, deep-veined leaves. Chinese cabbages, such as bok choy and 'Wong Bok,' do not form tight heads. The leaves are loose and can be harvested a few at a time or all at once. If leaves are taken in this "cut and come again" method, the lower leaves should be harvested first.

Cultivation

Cabbages are well suited for fall and late-winter plantings in Texas. They can be planted in the spring, but spring temperatures generally heat up too fast and the plants will bolt rather than produce mature heads. They do best if container grown and set out in pre-moistened soils when they are 4 to 5 weeks old with five to six true leaves. Cabbage plants are more cold tolerant than other crucifers and mature best when temperatures are between 60°F and 65°F. Most cabbages are ready for harvest in 60 to 90 days after planting. Consistent watering is important; however, cabbages can split if they receive too much fertilizer or water. Spread 2 to 3 inches of mulch around the plants, being careful that it is not piled up against the stem or covering the leaves.

VARIETIES

Green Cabbages

Bacalan de Rennes Flavorful, pointed heads that mature early. Good for spring planting.

Brunswick Very good freeze resistance. Stores well.

Couer de Boeuf Flavorful, pointed heads that mature early. Good for spring planting.

Danish Ballhead Forms large, blue-green, round heads, weighing up to 6 pounds. Resists bolting and splitting. Can be susceptible to fusarium yellows. Stores well.

Early Flat Dutch Large, flat heads, weighing 6 to 8 pounds. More heat resistant than other green cabbages. Stores well.

Early Jersey Wakefield Small, pointed heads, weighing up to 2 to 4 pounds. Good resistance to cabbage fusarium yellows. Compact plants can be grown closer together.

Henderson's Charleston Wakefield Larger than 'Early Jersey Wakefield,' weighing 4 to 6 pounds. Good variety for hotter climates.

Late Flat Dutch Very large blue-green heads that can grow to over 10 pounds. Stores well.

Red Cabbages

Mammoth Red Rock Also known as 'Red Danish.' Large, deep red heads that have good flavor and color.

Red Acre Small, compact heads that weigh about 3 pounds. Resistant to fusarium yellows. Stores well.

Ruby Perfection Deep red, solid heads that are sweet and resistant to cracking. Heavy yield.

Tete Noire Solid red heads. French variety.

Savoy Cabbages

Chieftain Savoy Flattened, round heads, weighing 6 to 8 pounds. Tolerant of more extreme high and low temperatures.

Savoy Drumhead Perfection Gray-green, deeply savoyed heads that can weigh 6 to 8 pounds.

Chinese Cabbage

Bok Choy White to light green stalks, 8 to 12 inches long. Slow to bolt.

Gai Lohn Also referred to as 'Chinese Kale.' Silver-green variety, with stalks 8 to 10 inches long. Edible broccoli-type flowers, young leaves, and stalks. More heat tolerant than other greens.

Komatsuna Large, oval leaves. Somewhat tangy flavor. Good as a "cut and come again" planting.

Michihili Chinese Loose-leaf cabbage with dark green leaves on white stems, 14 to 24 inches in height.

Shanghai More heat-resistant pak choi variety.

Tah Tsai Deep green, glossy, oval leaves look similar to spoons. Forms winter rosettes. Easy-to-grow pak choi variety with good heat and cold tolerance.

Wong Bok Loose-leaf cabbage that forms a head 9½ inches tall and 6½ inches wide. Light green leaves and white stems.

CAULIFLOWER
Brassica oleracea var. *botrytis*

✳ **DESIGN USES:** Like the other edible crucifers, cauliflower is a dramatic addition to the garden. The snowy white heads form at the center of gray-green leaf whorls. Cauliflower plants make wonderful accents or can be scattered among other plantings for dramatic punctuation. Purple-heading varieties are especially dramatic.

Description
Cauliflower is a temperamental grower in Texas, preferring very specific temperatures. It must

Cauliflower (Photo by Krishnan S. Navaneeth)

mature before temperatures reach 75°F and can be harmed by severe cold in the winter. Many times successful cauliflower growing is just a matter of the whims of Texas weather. It is helpful to plant early-maturing varieties.

Cultivation
Cauliflower can be planted in spring or fall; however, it performs best if planted in the fall in Texas. Transplants should be set in the garden 8 to 10 weeks before the average first frost date in the fall and 2 weeks before the last anticipated frost date in the spring. In areas of Texas where freezing weather is not an issue, cauliflower can be grown through the winter. Space the plants 18 to 24 inches apart. At the first sign of the forming heads, or "curds," the

outer leaves should be pulled together to cover the heads, unless they are self-blanching, and left tied until harvest. Maintain consistent watering, and spread 2 to 3 inches of mulch around the plants. Do not allow mulch to lie on the stalks or leaves.

Varieties

Broccoverde Green heads, weighing about 1 pound, that retain their color when cooked. Good flavor. More heat tolerant than other varieties.

Early Snowball Early producing (60 days) with 6- to 7-inch heads.

Giant of Naples Vigorous plants with large heads weighing up to 3 pounds.

Green Macerata Bright green heads weighing 2 pounds. Good raw or cooked. Vigorous plants.

Purple of Sicily Bright purple heads weighing 2 to 3 pounds. Turn bright green when cooked. Some insect resistance.

Snowball Self-Blanching Leaves naturally wrap over the heads, unless heads grow larger than 6 inches in diameter.

Violetta Italia Large plants with bright purple heads that turn bright green when cooked.

GREENS
Brassica spp.

❋ **DESIGN USES:** These beautiful, nutritious plants can add interesting textures and colors to cool-weather gardens. A variety of foliage colors from blue-green to purple-red with varying degrees of contrasting vein colors afford these plants a prominent place in the garden. Use greens as contrast, background, or accent plants.

Description
Three main types of leafy greens in the mustard family are grown in Texas: collard greens (*B. oleracea* var. *acephala*), kale (*B. oleracea* var. *acephala*), and mustard greens (*B. juncea*). Greens in the mustard family are known for their bold flavors, which some people may find a little overwhelming. Greens may have a milder flavor if harvested after a frost. Greens can be planted in the fall or early spring, seeded directly in the garden or from container-grown plants.

Collard Green Varieties

Even Star Land Race Small leafed with a mild flavor. Tolerant of powdery mildew.

Georgia Southern Good producer of wide, blue-green leaves. Good heat and freeze tolerance.

Green Glaze Lemon-green, glossy, somewhat waxy leaves with a bold texture. Grow to about 1½ feet. Waxy coating provides some natural protection against cabbage loopers. Resistant to heat and cold and slow to bolt.

Morris Heading Loose heads of flavorful dark green leaves with light green veins. Fast growing and slow to bolt.

White Mountain Cabbage A very large plant (up to 3 feet) with dark green, savoyed leaves. Forms a loose head, like 'Morris Heading,' but is much larger.

Kale Varieties

Dwarf Blue Scotch A dramatic plant about 15 inches tall and 2 feet wide with bold-textured, blue-green, ruffled leaves.

Russian Red Mild-flavored, red-tinged, oak-type leaves with reddish stems.

Savoy Cross Cross between 'Dwarf Scotch' and 'Black Tuscan.' Vigorous growth with dark green, purple-tinged, slightly ruffled leaves.

Toscano Dark gray-green, heavily savoyed, loose leaves that can reach 2 feet in length. Good flavor. Good heat and cold tolerance.

Mustard Green Varieties

Green Wave Upright habit with broad, ruffled, green leaves. Spicy flavor.

Osaka Purple Red-purple, slightly ruffled leaves on white stems. Spicy flavor.

Red Giant Bold maroon color that makes a beautiful garden accent. Large, oval, slightly ruffled leaves.

Ruby Streak Deep red, narrow, deeply lobed leaves on green stems. Overall texture delicate and feathery. Leaf shape adds drama to the landscape.

Savoyed kale (Photo by Andrew Edmonson)

'Red Giant' mustard with lettuce and strawberries

Southern Giant Curled Light green leaves with ruffled edges, with a height of 18 to 24 inches.

Tendergreen Smooth, dark green leaves with a milder flavor than that of many types of mustard.

Tokyo Bekana Small, oval, light green, slightly ruffled leaves. Fast growing with a mild taste.

KOHLRABI
Brassica oleracea var. *gongylodes*

Description
Kohlrabi is a very odd-looking vegetable. The plant forms a bulblike structure, a swollen stem above the soil line. Leaf stems grow from the swollen stem that have large, gray-green leaves. This whimsical vegetable is a very good choice for Texas gardens and is tasty and nutritious.

Cultivation
Kohlrabi is relatively easy to grow as long as it is provided with well-drained, fertile soil and consistent watering. The stem will become woody if allowed to dry out. Kohlrabi is seeded directly into the garden bed and can be sown in spring or fall. Spring plantings should begin when evening temperatures are consistently in the 40s, and fall plantings should be timed to allow 50 days before possible temperatures in the mid-20s. Some areas of Texas have temperatures above the 20s year-round, so kohlrabi can be grown in these areas all winter; however, temperature fluctuations may cause the plants to go to seed. Plant seeds ½ inch deep in well-moistened beds, and firm the seeds in the bed to achieve good contact between the soil and the seed. The seeds can be clumped every 6 inches with three to four seeds per hole. Thin the plants to 6-inch spacing to allow room for the plants to mature.

Kohlrabi (Photo by H. Zell)

Varieties

Early White Vienna Light green bulbs about 2 inches in diameter with creamy flesh. Flavor somewhat tangy like that a turnip. More compact than 'Purple Vienna.'

Purple Vienna Purple color of bulb and stem make this a very striking addition to the garden. Harvest the mild-flavored bulbs when about 2 to 2½ inches in diameter before they become woody.

RADISH
Raphanus sativus

❋ **DESIGN USES:** Radish plants are not particularly showy in the garden. The edible part forms below the ground, and the foliage is not particularly interesting. Radishes make a nice addition when planted in a border or as little pocket plantings, tucked in with other plants.

Radish plants (Photo by Andrew Edmonson)

Description

With proper growing conditions, radishes are very easily grown Texas. The leafy part of the plant is most noticeable during the majority of the plant's growth. The "root" will bulb up as it matures, showing a variety of red, pink, and cream-colored shoulders above the soil line. The fleshy part of the plant generally thought of as a root is actually a hypocotyl, or enlarged stem, that grows above the roots.

Two types of radishes are grown in Texas gardens. The daikon varieties are larger and generally planted in the fall. The smaller varieties are more commonly planted in the spring. The plant is usually grown for the spicy root, but the leafy greens are edible as well and especially flavorful in certain varieties.

Cultivation

Radishes are easily grown if a few simple rules are followed. First, add plenty of organic matter to create a loose, friable soil that is rich in nutrients to encourage lush top growth and ease of root formation. Second, water consistently to keep the soil from drying out. Radishes can split from uneven watering. Third, and perhaps most important, plant at the right time of year. Seed radishes directly in pre-moistened beds when soil temperatures are above 45°F for spring planting and below 80°F for fall planting. Spring planting can continue until 30 days before temperatures are expected to reach 80°F. Fall planting can be extended until 60 days before nighttime temperatures are expected to fall into the 20s. Because radishes mature quickly, they can become overcrowded if not thinned soon after they emerge. Thin the smaller spring varieties to 2 inches apart and the larger fall-planted varieties to 4 to 8 inches apart. Like other brassicas, radish plants are biennial and will bolt if exposed to alternating cold and warm temperatures. The roots become spicier as they mature, and after they begin to produce seeds, they may become too tough to eat. Harvest when the "shoulders" appear above the ground.

Daikon Radish Varieties

Long Black Spanish Strongly flavored black radish with white flesh that grows to about 9 inches long. Baking mellows the flavor. Overwinters well.

Misato Green Variety with green flesh that grows to 8 inches in length. Overwinters well.

Misato Rose Pink and light green skins with rose-colored flesh. Best eaten when about 4 inches long. Colorful variety for pickling.

Round Black Spanish Similar to 'Long Black Spanish' except in size. Round roots 3 to 4 inches in diameter.

Tokinashi White, flat radish that is 1 inch long and 3 inches wide. Strong flavor.

Radish Varieties

Champion Bright red with a mild flavor.

Cherry Belle Red radish with crystal-white flesh and mild flavor.

Crimson Giant Grows from ½ to 2 inches in size and maintains mild flavor. Some heat tolerance.

Early French Breakfast Bright red with white tips and full flavor.

Easter Egg Multicolored mixture of pink, lavender, red, and white.

Lady Slipper Bright pink, elongated radish with a blunt tip. Mild flavor.

White Icicle White radish 4 to 5 inches long. Mild flavor and some heat tolerance.

TURNIP
Brassica rapa var. *rapa*

❋ **DESIGN USES:** Turnip plants have broad, elliptical leaves that emerge from the base of the plant. The plants grow about a foot in height and can be attractive when planted in little groupings, interspersed with other plants.

Description

Like radishes, the "shoulders" of the root appear below the soil line as the plant matures, showing a light cream-colored root with purple markings. Both the roots and the greens of the plant are edible, and some varieties are cultivated specifically for their greens.

Cultivation

Turnips do very well in Texas gardens if planted in loose, friable soils with plenty of well-decomposed organic matter and consistent watering. They are generally planted in the fall when temperatures are below 80°F and will mature best when temperatures are between 40°F and 80°F. Planting can continue until 6 weeks before temperatures are expected to fall into the 20s. Turnips can be planted in the spring as well, but the window for planting is more limited. Seed turnips when temperatures are above the 20s, and discontinue 6 weeks before daytime temperatures are expected to be in the 80s. Plant seeds in pre-moistened soil, and thin seedlings to 2 inches apart if growing for the greens only. For proper root formation, thin the seedlings to 4 inches apart. Turnips germinate and grow quickly under ideal conditions so they will need to be thinned soon after they emerge. Thinning should occur within the first 2 weeks. Young turnips have a milder flavor than fully mature ones. Turnips overwinter well, especially if they are mulched.

Varieties

Amber Globe Up to 6 inches in diameter, but milder if harvested at 3 to 4 inches. Creamy yellow flesh.

Ideal Purple Top Milan White with purple tops and a flat shape. Mild flavor.

Navet des Vertus Marteau Very tender, cylindrical, white roots, growing 5 to 6 inches in length.

Purple Top White Globe White with purple shoulders. Mild flavor.

Seven Tops Grown especially for the greens, not the woody roots.

Shogoin Japanese variety grown for its roots and greens. White roots have mild flavor. Can tolerate some hot weather.

Tokyo Market All-white turnip with a mild flavor that grows 4 to 5 inches long. Good for pickling when young.

Cucurbits

Description

Members of the Cucurbitaceae family are a challenge for gardeners in Texas. Cucumbers, melons, and squash hold such promise when they are first planted. Their big, bold leaves, bright yellow flowers, and rapidly growing vines seem to defy any insect or disease predators. Unfortunately, many hard-to-control insects and diseases accept the challenge. For the dedicated gardener, these beautiful plants can be very rewarding. They just require more effort and vigilance than many other edible plants. The vines need to be inspected daily for diseases, as well as insect damage and eggs. Bats are voracious feeders of harmful insects, and it would be wise to put up a couple of bat houses for insect control if possible. It is also a good idea to make successive plantings of cucurbits, allowing various stages of the plant to act as "catch crops"

Squash blossoms (Photo by James Beesley)

be given plenty of space in the garden. Squash and melon vines can rapidly consume a small garden. These vines grow very rapidly, and their roots reach as wide as the vines do.

Cultivation

Plant cucurbits in full sun in raised beds or very well-prepared garden patches. Cucurbits are heavy feeders and will appreciate deeply worked soils. Prepare the soil prior to planting with plenty of well-decomposed compost in an area as large as the plants will consume. Space usage will vary according to species and variety; however, it is best to assume that these rambunctious vines will occupy more space than they have been allotted. It is also a safe bet that the average gardener would not overdo the depth and breadth of organic matter that should be worked into the bed, and the plants will certainly respond to the effort.

After the beds have been adequately prepared and the soil temperatures have warmed to between 60°F and 70°F, sow the seeds in clusters of three or four and thinned to one or two plants per cluster after germination. Space clusters 6 inches apart. Although it is a common practice to set out seedlings, cucurbits germinate quickly and there is no real benefit from setting out transplants.

Consistent watering is very important to cucurbits. The soil should not be allowed to completely dry out, which results in misshapen or improper fruit formation. Watering should be done from a drip system or a hose placed at the base of the plant to prevent water from getting on the leaves. Cucurbits are very susceptible to powdery mil-

for insect infestations as they occur. Be sure to destroy infested plants rather than compost them.

From a design perspective, cucurbits can be very effective when grown as bold-textured ground covers or as vertical elements. These plants have tendrils that will cling to supports and can be grown on trellises or tepees. Larger fruits, such as melons and large squashes like pumpkins, will need to be supported with slings to prevent the vines from being pulled away from the supports. Some gardeners use old stockings for this purpose. Cucurbits like to spread out in the garden. If the vines are left to grow on the ground, they should dew, a fungal disease spread by moisture. After they develop their first true leaves, water with fish emulsion and seaweed and continue applications every week to 10 days for the first month. After this a foliar application should be fine. Layer 2 to 3 inches of mulch around the plants after they are a few inches tall. Because of its fluffy texture and slow decomposition, pine straw is an especially good choice around ground-trailing vines to keep the fruit from coming in contact with the soil.

Members of the cucurbit family have male and female flowers on the same plant and require a pollinator to spread the pollen from flower to flower. Bees are the primary natural pollinator for cucurbits, and a lack of a local bee population can mean no squash for the dinner table. The gardener can help by hand-pollinating. First, identify the female flowers by the swollen ovary below the blossom, which will become the future fruit. The male flower produces a stamen (anther) that contains the pollen. To hand-pollinate, pick the male flowers and strip away the petals; then brush the pollen from the stamens onto the female stigma inside the flower. The transfer can also be done with a cotton swab. Cucurbit flowers are open for only a short time, mostly in the morning. Once the flower has closed, the opportunity for pollination has passed.

CUCUMBER
Cucumis sativus

❂ DESIGN USES: Most cucumber varieties need to be grown on trellises or other upright structures. Bush-type varieties are pretty as accents or in groupings. Try planting them in a bed of annual color. The bold-textured foliage and interesting cucumber fruit are very pretty when used as a background for lower-growing plants, and they do not require as much ground space as their melon and squash relatives. Living walls and tunnels can be created with these beautiful vines. The yellow flowers are smaller than squash blossoms.

'Japanese Long' cucumber on bamboo support (Photo by Andrew Edmonson)

Description

There are two main types of cucumbers, slicing and pickling. The pickling varieties are usually smaller (about 3 to 4 inches long) and generally bumpier than the slicing varieties, which are usually around 6 to 8 inches long. The slicing varieties have been cultivated for fresh eating because they turn soft when pickled. The pickling varieties can be used for fresh eating as well as pickling. Cucumbers, like other cucurbits, are vining plants and respond well to training on trellises, or they can be grown in cages like tomatoes. This gets the vines off the ground and provides better air and light circulation around the leaves, making them less susceptible to fungal diseases like powdery mildew.

Cultivation

Cucumbers grow best in soils with a pH between 6.0 and 7.0. Adjust planting beds as necessary. A wide area of deep, well-prepared soil with plenty of organic matter is essential for success. Cucumbers respond well to raised beds and trellising and make a very pretty vertical element in the garden. Whether grown on trellises or on the ground, seedlings should be thinned to 2 to 3 feet apart to give the individual plants plenty of room for their extensive root systems.

Slicing Varieties

Chinese Yellow Very productive variety with yellow-orange skin when mature. Crisp, flavorful fruits about 10 inches long.

Edmonson Cream to light green cucumbers about 4 inches long. Sweet and mild flavor.

Heavy yields. Good for fresh eating and for pickling. Good disease and insect resistance.

Hmong Red Fruits begin white to pale green and turn golden-orange when ripe. Very productive, mild-tasting variety draws attention when vines are covered with green and red fruits.

Japanese Climbing Long, dark green fruits 9 inches long. Good for slicing fresh or pickling.

Japanese Long Very productive, crisp variety with a mild flavor. Fruits long and slender. Firm flesh with fewer seeds than many other varieties.

Lemon A semi-bush type that produces light green fruits similar in appearance to lemons. Best if picked before they turn yellow.

Long Green Improved Long, medium-green fruits with black spines, 10 to 12 inches long, produced on vigorous vines. Can be eaten fresh or pickled.

Poona Kheera Potato-shaped Indian variety with a cream to light green skin that matures to light brown. Crisp and mild in flavor. Vines produce early with heavy yields. Reported to have some disease resistance.

Richmond Green Apple Light green fruits about the size of a lemon. Mild, sweet taste.

Straight Eight Vigorous vines that produce dark green 7- to 8-inch fruits. Some resistance to cucumber mosaic virus.

Suyo Brocade Asian variety that produces fruits about 15 inches long and about 1½ inches in diameter. Thin, ridged skin and flesh that is rarely bitter. Well adapted to hot Texas summers.

Thai Green A medium-green cucumber that grows to about 7 inches in length and about 2 inches in diameter. Well adapted to hot summers.

Uzbekski Fat cucumber that grows 6 to 8 inches in length. Good for both fresh eating and pickling. Stores well.

White Wonder White-skinned variety good for fresh eating and pickling. Grows 6 to 7 inches long and 2 to 3 inches in diameter with blunt ends. Good heat tolerance.

Yamato Asian variety that produces green fruit with yellow stripes and white spines. Grows 12 to 16 inches long. Similar to 'Suyo Long' with superior flavor and yield. Like 'Suyo Long,' well adapted to hot summers.

Yard Long Armenian A great conversation starter in the garden. Fruits can reach 3 feet in length but sweeter if harvested when about a foot long. Thin, pale green skins with ribs. Very good flavor. Tolerates heat.

Yok Kao Yellow-green, crisp fruits 5 inches long. Good flavor. Can be eaten fresh or pickled.

Pickling Varieties

Boston Pickling Dark green fruits 5 to 6 inches long with smooth skin and black spines.

Chicago Pickling Medium-green with blunt ends and bumpy skin. Grows 5 to 6 inches long. Very prolific with good disease resistance.

De Bourbonne Tiny, 2-inch-long cucumbers used to make cornichon pickles. Vines are good producers.

Fin de Meaux Dark green, 2-inch-long cucumbers used to make cornichon pickles. Vines are good producers.

Mexican Sour Gherkin (*Melothria scabra*) Tiny cucumbers (about an inch long) that look like little watermelons. Very decorative, prolific vines with smaller leaves than those of most cucumbers. Cucumber flavor a sour, lemony aftertaste. Can be eaten fresh.

Parisian Pickling A small gherkin or cornichon-type pickling cucumber.

SMR 58 Vigorous, productive vines. Resistant to scab and cucumber mosaic virus.

West India Burr Gherkin (*Cucumis anguria*) Not a true cucumber but used in much the same way as pickling cucumbers. Roundish, small, green fruits with dull, spiny points on the outside. Can be eaten fresh or cooked like squash. Very prolific plants well adapted to hot weather.

MELON (MUSKMELON AND WATERMELON)

Cucumis melo and *Citrullus lanatus*

✺ **DESIGN USES:** Melons require a lot of garden space. Both muskmelons and watermelons are commonly left to grow along the ground; however, they can be trellised if the fruit is attached to the trellis in a sling. Large varieties of watermelons will probably not be suitable for trellising. Melon

Charantais' melon (Photo by James Beesley)

vines can be used as a sprawling ground cover. Be careful to keep the vines from tangling with other plants. The yellow, buttercup-shaped flowers are not as showy as squash blossoms.

Description
Muskmelon fruit includes casaba melons, honey-dew melons, and orange-fleshed melon varieties referred to as "cantaloupe," although true can-taloupes are actually a different fruit from what is customarily sold as cantaloupe in American markets.

Cultivation
Melons prefer a soil pH from 5.5 to 7.0. They are very well suited to Texas soils if they are given plenty of organic matter. They will not grow in heavy or poorly drained soils. To produce large fruit, melon vines require consistent moisture and plenty of sun. Muskmelons will need at least a 6-foot-diameter circle from the center of the hill, and watermelons will need an additional foot or two in diameter as growing space. Watermelon vines can be expected to sprawl widely. Be pre-pared for a 24-square-foot area for a single vine. Trellising can help conserve space. If vines are not trellised, it is a good idea to put dry mulch, such as pine straw, under the fruit to prevent soilborne fungal diseases from rotting the fruit.

Orange-fleshed muskmelons are ready for har-vest when the stem separates easily from the vine or when the skin between the netting turns from green to yellow. Green-fleshed varieties are ready when they soften slightly on the end and change slightly in color. Watermelons are ripe when the belly side turns light yellow or cream colored and the tendril where the vine is attached to the fruit starts to wither. Be cautious when turning the fruits for inspection.

Muskmelon Varieties: Orange Fleshed
Ananas d'Amerique a Chair Verte Ananas means "pineapple" in French, and this variety has a slight pineapple taste. Flesh is green

closer to the rind and orange near the center. Fruits usually weigh about 1½ pounds.

Banana A unique-looking melon with slender, pale yellow fruit 14 to 18 inches long. Salmon-yellow flesh that is soft and juicy.

Charentais A French variety that is a true cantaloupe. Small fruit about the size of a softball, with smooth, light green skin with darker green, symmetrical, longitudinal stripes. Very sweet, fragrant fruit without the "muskiness" of muskmelons.

Crane (**Eel River**) A cross of several varieties that have produced a 3- to 5-pound melon with a unique "teardrop" shape. Light orange flesh that is sweet and fragrant.

Crenshaw Large, oval melon with very good flavor. Yellow-green skin and pink-salmon flesh. Grows well in warm, dry climates.

Edisto 47 Melons that weigh 4½ pounds and have flavorful, salmon-colored flesh. Well adapted to hot, humid conditions.

Gaucho Grows to about 14 by 8 inches. Rinds are golden yellow with deep ribs. Very juicy and aromatic, creamy yellow flesh .

Hale's Best Heirloom variety weighing 3 to 5 pounds with sweet orange flesh.

Hearts of Gold Dark green skin and deep orange flesh, weighing 3 to 4 pounds. Sweet taste and fragrance.

Honey Rock (**Sugar Rock**) Ribbed, gray-green skin that turns a creamy yellow when mature. Salmon-red, sweet flesh. Vigorous vines that produce 3- to 4-pound fruits. Some resistance to fusarium wilt.

Noir de Carmes A true cantaloupe. Deeply ridged, very dark green skin that ripens to orange-yellow. Light orange, sweet, fragrant flesh. Fruits relatively small, 3 to 4 inches. Good for trellising.

Persian Large, 6-pound melons with bright orange, sweet flesh.

Prescott Fond Blanc Large, 4- to 9-pound melons. Deeply ribbed, gray-green skin that turns a light yellow when ripe. Salmon-colored, very sweet, fragrant flesh.

Sierra Gold Grows to 3 pounds. Thick, salmon-colored flesh. Some resistance to powdery mildew.

Sweet Passion Sweet melons that grow to 3 to 4 pounds. Some drought and wilt tolerance.

Muskmelon Varieties: Green Fleshed

Boule d'Or A very sweet melon with green flesh and golden-yellow skin. Large, somewhat flattened fruits, typically 7 by 9 inches and weighing 6½ pounds.

Early Hanover Sweet-fleshed melon that grows to 2 to 3 pounds.

Eden's Gem Softball-sized melon with sweet, spicy flavor. Good size for trellis growing.

Green Machine Very prolific vines that produce 2-pound, sweet melons.

Green Nutmeg Flattened, oval shape with yellowish-brown skin flecked with green. Weighs up to 2 pounds. Sweet, spicy taste. Early producer.

Honeydew Green Oval, 5- to 7-pound fruits that turn from pale green to creamy yellow when mature. Sweet flavor. Resistant to heat so a good choice for Texas.

Jenny Lind Turban-shaped, 1- to 2-pound melons with light green, very sweet flesh.

Valencia Winter Melon Does not grow in the winter. An Italian variety that may be named for its ability to keep for up to 4 months. Dark green skin and cream-colored flesh.

Vert Grimpant (**Green Climbing**) Smaller fruit (1 to 2 pounds) that is a good choice for trellising.

Muskmelon Varieties: Casaba Melon

Casaba Golden Beauty Large fruits, typically 7 to 8 pounds. Wrinkled skin turns yellow when ripe. White flesh with a sweet smell. Well adapted to hot weather. Stores well.

Watermelon Varieties

Ali Baba Light green melon weighing 15 to 25 pounds, produced on vigorous vines.

Black Diamond Dark green rind with sweet, red or yellow flesh. Grows to 16 to 18 inches long. Late season.

Black Tail Mountain Round, small (9-inch diameter) fruits on very productive vines. Dark green rind with darker stripes and sweet, orange-red flesh.

Charleston Gray Gray-green rind and red flesh with good consistency and sweetness. Grows to 20 to 40 pounds.

Congo Very large, 30- to 40-pound melon with a thick, striped rind. Deep red, very sweet flesh.

Crimson Sweet Very large, up to 25 pounds,

and can produce numerous sweet fruits per vine. Some resistance to anthracnose and wilt.

Daisy Also called 'Yellow Shipper.' Bright yellow, sweet meat. Grows to between 13 and 20 pounds and has a thick rind.

Desert King Light green rind with sweet, yellow flesh. Resists sunburn and handles drier conditions than many watermelon varieties. Stores well.

Dixie Queen Large, round variety that can grow to 50 pounds. Striped, dark green and ivory rind with bright red, sweet flesh.

Golden Midget Miniature melon that grows to about 3 pounds. Golden-yellow rind when ripe. Salmon-colored, sweet flesh.

Jubilee Large, long, oval-shaped melon with deep red, crisp flesh.

Malali Small, sweet melon from Israel . Green-striped rind with light red flesh. Fruits reach about 10 pounds. Well adapted to hot climates.

Moon and Stars Interesting-looking watermelon, with various-sized yellow blotches (large ones like "moons" and small ones like "stars") on a dark green rind. Very large, red-meated variety that can grow to 40 pounds. Leaves have yellow speckles. A yellow-meated variety available called 'Moon and Stars Yellow Fleshed.'

Orangeglo Deep orange flesh that is exceptionally sweet. Yields high. Strong vines resistant to wilt and some insects.

Quetzali Pink-fleshed fruit that is very sweet with a good texture. Very thick rind with alternating light and dark stripes.

Strawberry Dark green rind, ½ inch thick, with darker green stripes. Sweet red meat. Grows to about 8 by 20 inches and weighs 15 to 20 pounds. Good disease resistance.

Sugar Baby Dark green rinds when ripe. Grows to 6 to 8 inches long. Consistently produces 6- to 10-pound fruits with small seeds.

Sugarlee Sweet, pink-red flesh and striped rinds, weighing 15 to 18 pounds. Good disease resistance.

Verona Large, round watermelons with dark green rind. Sweet, crisp, red flesh with good flavor. Resistance to disease. Stores well.

White Wonder Small, 3- to 10-pound "icebox"-sized melon. Creamy white, sweet flesh. Green rind with darker green stripes.

SQUASH
Cucurbita maxima, *C. moschata*, and *C. pepo*

❋ **DESIGN USES:** Squash plants have large, bold-textured leaves and, large, golden-yellow flowers, providing a beautiful summer display, especially when mixed with other colorful flowers like cosmos and zinnias. Vining types will require a large area if left to grow along the ground. Squash can be trellised if space is limited. Bush-type varieties make nice accents and groupings in the midrange scape.

Description
Several species of squash can be grown in Texas. The *C. maxima* species includes hubbard and buttercup squashes as well as some larger pumpkins. *C. moschata* is butternut squash, and *C. pepo* includes acorn and summer squashes like zucchini and yellow crookneck, as well as most pumpkin varieties. Squashes within the same species will cross-pollinate.

Squashes are further classified into winter and summer types, which refers to storage qualities rather than the season when they are grown. Summer squashes are harvested when the skins are still soft, and winter squashes are harvested when the skins are hard, so they store better. The flowers are also edible and can be used as a bold accent on a platter, or they can be stuffed or deep-fried.

Cultivation
Like all cucurbits in this section, squash plants need a large growing area. Sow seeds in moistened, well-prepared beds when all danger of frost has passed and the soil has warmed to at least 60°F. Seedlings usually emerge in about 5 to 7 days and should be thinned to 3 feet apart and given 3 to 8 square feet of growing area, depending on variety. The bush varieties will do well in a 3-foot-diameter growing area; vining types will need 6 to 8 square feet each. Thinning time is also a good time to apply a 2- to 3-inch layer of mulch.

Summer squash matures quickly after pollination, some varieties in only 3 to 6 days. Squash vines should be inspected daily for insect eggs on the underside of the leaves. At the same time, check regularly for ripe squash. Winter squashes take much longer to mature, usually 60 to 80 days. Winter squash can be stored in a dry medium at temperatures between 50°F and 65°F for up to 6 months.

Squash plant

Summer Squash Varieties

Bennings Green Tint Scallop A bush type that produces light green, 3-inch, saucer-shaped squash. Pale green flesh with good flavor, especially when picked young.

Cocozelle Bush (**Costata Romanesco**) A bush type with Italian origins that produces 10-inch, slender, striped zucchini.

Greenish-white flesh that retains its flavor even when harvested at full maturity, but flavor better if harvested when young.

Early Prolific Yellow Straightneck A bush type that produces heavy yields of golden-yellow, club-shaped fruit. Flavor best if fruits harvested when young.

Early White Bush Scallop A bush type that

produces 4- to 7-inch, smooth, flattened, white-green squash with scalloped edges. Fine-grained, white flesh with a sweet flavor.

Lagenaria Longissima Produces long, light green squashes that are interesting hanging from a trellis. Resistant to squash vine borer.

Tatume Oval, yellow-green squash. Grows on vigorous vines that need plenty of room. More drought tolerant than many squash varieties. Some resistance to squash vine borer because the stems more solid than those of other squashes. Flavor sweeter when cooked.

Tinda Gourd More closely related to watermelon. A variety from India used like squash in stir-fries and curries. Harvest fruits when about 3 to 4 ounces.

Tromboncino Long, somewhat curvy, light green squash with a bulb shape on the lower end. Very interesting hanging from a trellis.

Yellow Bush Scallop A bush type that produces 7-inch wide, scalloped, yellow squash on vigorous vines. Some resistance to squash bugs.

Zucchini Black A bush type that produces 6- to 8-inch fruits on vigorous plants. Very deep green to almost black rinds with greenish-white flesh.

Winter Squash Varieties

Black Fatsu Black Japanese variety that turns chestnut-brown in storage. Flattened, round, heavily ribbed fruits with nutty-flavored, golden flesh. Prolific vines that produce 3- to 8-pound fruits. Some insect resistance.

Blue Hubbard Large squash, about 12 pounds, with blue-gray rind and deep gold flesh. Sometimes planted as a "catch crop" to lure cucumber beetles from other cucurbits. Green and golden varieties also available.

Buttercup A vining variety that weighs about 3 pounds. Green-skinned squash with sweet, dry, orange flesh. A bush variety also available.

Cornell's Bush Delicata Relatively small plants with a 3-foot spread, but high yielding. Fruits about 8 inches with sweet, gold flesh. Resistant to powdery mildew.

Delicata A vining variety that produces 7- to 9-inch fruits with light green rinds that have darker green stripes. Deep orange, tender flesh with a somewhat sweet flavor.

Fordhook Acorn Tan-colored acorn squash that weighs about 2 pounds. Yellow flesh that can be fried like a zucchini when young.

Marina di Chioggia Dark blue-green, bumpy rind and smooth, orange flesh.

Ponca Butternut Compact vines, so a good choice for smaller areas, even containers. Produces 2-pound squash with a tan rind and orange flesh.

Queensland Blue Pretty, large squashes with a blue, deeply ribbed rind and deep orange flesh. Stores well.

Sibley Banana shaped with sweet orange flesh.

Spaghetti Squash Large, yellow-fleshed squash with stringy, pulplike flesh that can be used as a nutritious substitute for spaghetti pasta. Can be stored up to 6 months.

Sweet Meat A very large squash, reaching 15

to 20 pounds. Blue-green rind and orange flesh with a very good texture. Can be stored up to 6 months.

Table Queen Bush Acorn A vigorous vining variety that produces 2- to 3-pound green-skinned squash with yellow flesh. Deeply ribbed and round to ovoid. Stores well.

Thelma Sanders (Sweet Potato Squash) White acorn squash originally cultivated by Native Americans. Prolific vines produce fruits that weigh about 2 pounds with pale orange flesh.

Triamble Interesting 3-lobed fruits with blue-gray rinds that weigh 4 to 12 pounds. Thick, orange, sweet flesh.

Pumpkin Varieties

Although pumpkins are actually winter squashes, they are listed separately here because of their unique usage and numerous cultivars. Pumpkins to be carved as jack-o-lanterns should be planted around the Fourth of July to produce pumpkins for Halloween.

Americano Tonda An ornamental variety with an orange skin and green stripes. About 4 to 6 pounds.

Australian Butter Large, orange squash that grows to 15 pounds. Orange, thick, flavorful flesh. Good pie variety that stores well.

Baby Bear A miniature variety. Weighs 1½ pounds and is about 4 inches tall and 6 inches wide. Seeds do not have hulls and good for roasting.

Big Max Can grow to 100 pounds. Thick flesh good for pies and canning.

Boston Marrow Large pie pumpkin that grows 16 inches long by 12 inches wide. Deep reddish-orange skin with orange-yellow flesh. Traditionally grown by Native Americans. Shape similar to that of hubbard squash.

Cheyenne Bush Compact, bush-type plants that produce 5- to 8-pound pumpkins in a more limited space.

Cushaw Green Striped Crookneck Pear shaped with a curved or "crook" neck. Ivory skins with irregular green stripes. Pale yellow flesh that is somewhat coarse and sweet. Grows 16 to 20 inches long.

Galeux d'Eysines Light orange pumpkins with a lot of character. Covered with light yellow bumps or "warts." Round, flattened fruits that weigh 10 to 15 pounds. Deep orange, sweet, smooth flesh, making this pumpkin both ornamental and edible.

Jack Be Little Miniature pumpkin about 2 by 3 inches. Prolific vines. Fruits can be used for cooking or for decorations.

Jarrahdale Blue-gray skin makes this a very attractive addition to the garden. Flattened, deeply ribbed pumpkin that weighs 6 to 10 pounds.

King of Mammoth Very large pumpkin that can weigh as much as 250 pounds but more commonly 40 to 100 pounds. Skin irregular in color and varies from pinkish-orange to yellow.

Lakota Pear-shaped, very attractive variety.

Bright red with green streaking on the bottom. Nutty, sweet flesh. A traditional variety grown by the Lakota Sioux.

Musquee de Provence Beautiful, flattened pumpkin that is dusty orange and deeply ribbed. Turns orange-brown when ripe. Deep orange, thick, flavorful flesh. Weighs about 20 pounds.

Rouge Vif d'Etampes Very pretty "Cinderella" pumpkin. Deep orange, deeply ribbed, flat in shape, usually around 6 inches in height and 18 inches in diameter. Better for decoration than for eating.

Seminole Wild squash from the Everglades with tan, lightly ribbed skin and deep orange, sweet flesh. Weighs about 3 pounds. Some insect and disease resistance.

Small Sugar Grown for its superior pie quality. Small fruits that weigh 5 to 8 pounds. Yellow-orange, smooth, sweet flesh.

Winter Luxury Pie Lightly netted, flavorful pie pumpkin with a light orange, smooth rind and thick flesh. Weighs about 6 pounds. Very good for pies and pretty on the vine.

Legumes

Description

Members of the Leguminosae (also called Fabaceae) family include soybeans, beans, and peas. Vining varieties are very attractive upright elements when grown on trellises or tepees. These wonderful plants actually feed themselves and the soil by "fixing" nitrogen. In this process the plants pull nitrogen from the air, where it is plentiful, and put it into the soil through special roots that contain symbiotic bacteria called *Rhizobia* within nodules in their root systems. The first time legumes are planted in a particular area, they should be inoculated with *Rhizobia* bacteria. Legume inoculant can be purchased at most garden centers and nurseries. It does not keep from year to year, so it is important to make sure it is fresh. When legumes die, the nitrogen in their roots is released into the soil, making it available for future plantings. Because legumes have this unique ability to fix nitrogen, they make very good cover crops and should be rotated throughout the garden in various locations for this purpose.

Legumes are relatively care-free, although they are susceptible to some diseases and insects. Powdery mildew can be a serious problem, especially for peas, which are planted in the spring and fall when rainy weather can easily spread the fungal spores. All legumes will benefit from a drench of organic fungicide at planting time to help control soilborne fungal diseases.

Cultivation

Legumes should be planted in full sun in a spot where they have good air circulation. Peas and vining (pole) beans need to be trellised, and trellises should be in place before or at the time the seeds are planted. Sow seeds directly in raised beds or hills after they have been inoculated. Be careful that the soil does not become dry and crust over, as this will hinder the seeds' emergence. Water the seeds in with fish emulsion and seaweed at the

'Romano' and 'Yard-Long Asparagus' beans, Boggy
Creek Farm, Austin, Texas

time of planting, and continue applications every 7 to 10 days for the first month. Legumes are able to feed themselves, so they do not require much nitrogen from compost; however, beds should have plenty of organic matter to allow them to drain properly. Once the seedlings' leaves are clear of the ground, they will appreciate a good mulching.

BEANS

Phaseolus spp. and *Vigna* spp.

❋ **DESIGN USES:** Pole bean varieties will need to be grown on trellises or tepees. The fast-growing vines can be used to create living "walls" and "bean houses" for children's gardens with door and window openings. The scarlet runner bean is a particularly showy pole variety with red flowers that is grown both as an edible and ornamental plant. Purple-podded varieties can provide a dramatic contrast to other plants in the garden. Bush varieties can be grown as accents or groupings in a low to midrange scape.

Description

Beans are fairly easy to grow and produce an abundance of podded fruits that are a valuable protein source for people throughout the world. The two main types of beans are "pole" and "bush" beans. As the names imply, pole beans need to be trellised, and bush beans can grow without support. Bush beans are compact, typically growing about 12 to 18 inches tall. Pole varieties can typically grow 10 to 12 feet in length, and "runner"

'Chinese Red Noodle' beans (Photo by Rebecca Nickols)

beans can grow up to 20 feet. Bean color, size, and texture are quite varied.

Pole and bush beans are further categorized as "snap" and "shell." The pods of the snap types are eaten, and usually the seeds of the shell type are consumed. Shell beans may be left on the vines to dry before harvesting and then placed in a 175°F oven for 15 to 20 minutes to destroy any pests. The seeds should be completely dry before storing.

Cultivation

Beans grow best in soils with a pH between 6.0 and 7.5. Plant bean seeds directly into the soil after any danger of frost has passed. Beans create their own nitrogen, and an overabundance of compost will cause the plants to set more leaves than fruit. Beds should have enough organic matter to allow them to drain well. A healthy bean plant's root system will extend 3 to 4 feet into the ground and increase the bed's friability and fertility. The plants will appreciate deeply prepared beds, and raised beds are preferable.

Plant beans after soils have warmed to at least 60°F in pre-moistened, well-drained beds in full sun, 2 to 4 inches apart for bush beans and 4 to 6 inches apart for pole and runner beans. To extend the season, make additional plantings every 10 to 14 days until daytime temperatures exceed 75°F. The plants will not set flowers when temperatures are hot. Late-summer plantings should be timed to allow the flowers to set when daytime temperatures are below 75°F as well. Pole and runner beans are commonly planted on hills on tepee trellises. Plant four to six seeds per hill and thin to the strongest seedling. (Do not forget the inoculants.) Space the plantings 1½ to 2 feet apart. Large vines like scarlet runner will require more space. Water the seeds with fish emulsion and seaweed at planting time and again every 7 to 10 days for the first month.

Pole Bean Varieties

Plants in the *Vigna* genus listed here are related to southern peas and are well adapted to Texas climates. These species are included with pole beans because of their climbing habit.

Blue Coco Fleshy, slightly curved, bluish-purple pods, 6 to 7 inches long, with brown beans and purple-tinged leaves. Good heat tolerance.

Blue Lake Stringless Straight-podded, snap bean with white seeds and pods 6 to 7 inches long.

Cherokee Trail of Tears Prolific variety with a rich history. Said to be the bean that the Cherokee brought with them along the terrible march to Oklahoma known as the "Trail of Tears." Good as both a snap and a drying variety. Shiny, black beans.

Christmas Red Calico An 8-foot-long vine that produces a white lima bean with maroon splotches. Good variety for drying or can be eaten fresh. Flavor rich and nutty.

Dean's Purple Vivid purple bean pods that are very striking on a trellis against a background of green leaves. Good bean beetle resistance.

Florida Speckled Butter A lima bean with a long bearing season. Traditionally used as a planting between corn rows in the South. The 3-inch pods well adapted to heat.

French Horticultural A semi-runner bean. Vigorous vines that produce an abundance of 7-inch pods with light red streaking. Buff-colored beans with reddish blotches. Can be eaten fresh from the shell, frozen, canned, or dried for future use.

Helda-Romano A high-yielding Italian Romano type that grows to about 6 feet. The 6- to 10-inch pods are edible or can be shelled and eaten fresh, frozen, or canned.

Henderson's Black Valentine A snap bean that can also be dried. Very prolific.

Kentucky Wonder Vine grows to about 6 feet. Bean pods thick and grow 9 to 10 inches long. Very vigorous and resistant to rust. Beans can be eaten fresh, shelled, or with the pods or can be dried.

Kentucky Wonder Wax Vigorous vine grows 5 to 7 feet and produces heavy yields of yellow pods with light brown beans. Stringless if harvested when 5 to 6 inches long.

King of the Garden Very vigorous vine grows 8 to 10 feet long and produces over a long season. A lima bean variety that produces large beans in 5-inch pods.

McCaslan A stringless variety that can be eaten fresh or dried. Dark green, slightly flattened pods about 7 inches long. More drought tolerant than many varieties if beans harvested frequently.

Missouri Wonder Traditionally planted in cornfields with the cornstalks used as vine supports. Tolerates more extreme conditions.

Purple-Podded Pole Vigorous, 5- to 6-foot vines that produce 6- to 8-inch bluish-purple pods that turn green when cooked. Tan beans and light purple flowers. Heat tolerant.

Rattlesnake Vine grows to 10 feet and produces stringless, 7- to 8-inch dark green pods that are mottled with purple. The beans are tan with dark brown splotches.

Red-Noodle Asparagus *Vigna* genus. A long bean variety that produces 2-foot-long beans on 10-foot vines. A very dramatic garden accent. Tender, flavorful beans with no strings. Related to black-eyed peas. Very heat tolerant and stands up well to high humidity and insect pests.

Romano (**Italian Pole**) Prolific vines that produce flat-podded, flavorful Italian beans. Thick, stringless pods about 6 inches long. Good for canning, freezing, or fresh eating. A bush type available called 'Romano 14.'

Scarlet Runner A very beautiful, dramatic addition to the garden. Produces an abundance of scarlet-orange blossoms that become long, fleshy bean pods with purple-black beans. Both beans and flowers are edible.

Taiwan Black-Seeded Long Bean *Vigna* genus. Yard-long beans are a real conversation starter in the garden. Vines produce heavy yields of light green, 3-foot bean pods. Traditionally stir-fried rather than boiled.

White Half Runner A heat- and drought-tolerant bean good for canning, freezing, or shelling and eating fresh.

Yard-Long Asparagus *Vigna* genus. Produces a very dramatic, 20- to 30-inch bean pod. Flavor best if beans harvested just as the beans start to bulge in the pods. Loves the heat. A very striking red-podded variety available called 'Chinese Red Noodle.'

Bush Bean Varieties

Appaloosa A traditional bean of the Southwest that produces mottled bean pods in mixed colors of red, pink, purple, and green. Beans also mottled with splotches of red and brown.

Black Turtle A staple of indigenous Native American cooking in cultivation for thousands of years in the Americas. Good heat and disease resistance. Most often dried on the plant for dry beans, but young beans can be eaten green.

Blue Lake 274 A snap bean that produces heavy yields of 6-inch pods. White beans that fill out slowly and remain tender. Some resistance to bean mosaic virus.

Bolita A traditional variety from Mexico and New Mexico. Retains its rich flavor when dried.

Bountiful Produces heavy yields of green, 6-inch pods on 18-inch plants. Good for freezing or can be eaten fresh. Resistant to rust and mildew and some resistance to beetles.

Burpee Stringless Stringless bean pods produce waxy, dark brown beans. Heavy-yielding plant about 20 inches in height. Heat tolerant.

Bush Kentucky Wonder Bean pods thick and grow 9 to 10 inches in length. Like its pole bean counterpart, very vigorous and resistant to rust. Beans can be eaten fresh, shelled, or with the pods or can be dried.

Contender Grows to about 18 inches in height. Stringless pods about 6 inches long that remain flavorful even when larger. Resistant to viruses and heat tolerant.

Dixie Speckled Butter Peas Not a pea but a lima bean that produces small beans about the same size as peas.

French Flageolet The small green beans in French cassoulets. Shelled beans can be eaten fresh or dried. Beans jade green when fresh with white tinges when dried. Flavor at its peak when beans are fresh.

Landreth Stringless Early-producing variety with 5-inch stringless, tender, flavorful pods.

Pencil Pod Black Wax Pale yellow, 6-inch, tender, stringless bean pods. Producer over a long time. Heat tolerant.

Provider Early variety that produces heavy yields of 5- to 8-inch snap beans on 16- to 18-inch plants. Some resistance to mosaic virus.

Red Kidney A traditional variety from Mexico. Red beans in 6-inch pods that are good for drying, as well as for soups or baked beans.

Royalty Purple Pod Vigorous plants that produce an abundance of 5-inch, deep purple pods that turn green when cooked.

Santa Maria Pinquito A semi-bush variety that can be trellised, especially if overhead irrigation is used. Vigorous vines that produce a profusion of small pink pods. Tender-skinned beans with a creamy texture that holds up to cooking.

Tendergreen Shrub plants 18 inches tall that produce heavy yields of 6-inch, stringless pods. Heat tolerant and disease resistant.

Thorogreen Lima A productive lima bean that grows on 16-inch bushes. Bright green, 4-inch pods that produce three to five baby lima beans per pod. Can be eaten fresh, shelled, frozen, dried, or canned.

Soybean Varieties (Glycine max)

Agate A traditional variety from New Mexico that produces olive-green beans with a reddish tint on 1-foot tall, prolific plants.

Asmara A variety that grows to about 2 feet, bred for flavor and nutrition.

Beer Friend Named such because it is a favorite, and nutritious, snack when drinking beer in Japan. Bush-type plants grow to about 2½ feet and produce plump pods with three or four soybeans each.

Butterbeans An abundance of bright green pods produced on 2-foot-tall plants. Can be eaten fresh, canned, or frozen.

Moon Cake A taller variety than most, topping out at 5 to 6 feet, and produces large beans. Bred for flavor and nutrition.

PEAS
Pisum sativum

❋ **DESIGN USES:** Peas are highly susceptible to powdery mildew and are commonly trellised to keep them off the ground and allow air to circulate around the leaves. Some varieties are more upright and do not need to be trellised, but they will still need good air circulation. Delicate pea flowers in shades of lilac and white are very pretty in cottage gardens, mixed with lavenders, pinks, and soft yellows. Blue-podded varieties add an interesting element to cool-season gardens.

Description

Peas are grouped into three types: English peas, snow peas, and sugar snap peas. Additionally, they are classified by size and growth habit as dwarf, semi-dwarf, and vining. Peas are planted in the fall or spring in Texas. In areas of Texas with mild weather, winter plantings are possible. Peas are planted in the fall 8 to 10 weeks before the first frost date and are usually more successful than those planted in the spring. Peas need cool weather to mature, and spring weather tends to warm too quickly. Planting early-maturing varieties can help overcome this problem. Planting should cease when daytime temperatures are expected exceed 75°F. Pea plants will survive freezing weather, but the flowers can be damaged and the fruit will not set.

Cultivation

Peas grow best in soils with a pH between 6.5 and 7.5. East Texas gardeners may need to add lime to their soils. Like beans, peas do not require an abundance of nitrogen, since they produce it themselves. They will be fine as long as the soil is well drained and they have been inoculated with the *Rhizobia* bacteria. Planting in raised beds is a good idea to increase soil drainage and friability. Sow peas directly in the beds, 2 to 3 inches apart for dwarf and semi-dwarf varieties and 3 inches apart for vining peas. Layer 2 to 3 inches of mulch around the seedlings after they are tall enough to be above the mulch layer. Water the roots, not the leaves, consistently, especially during the bloom stage. Peas have a high sugar content and are particularly tasty when they are eaten soon after they are harvested, before the sugars turn to starch.

English Pea Varieties
Wrinkle-podded varieties are generally sweeter, and smooth-podded varieties are usually more winter hardy.

- **Alderman** (**Tall Telephone**) Vines grow to 5 feet tall. Pods about 5 inches long and contain eight peas per pod.
- **Blauwschokker** A versatile pea variety that produces purple flowers and purple-blue pods that are edible when harvested young. Mature peas can be shelled.
- **Green Arrow** A bush variety that grows 24 to 28 inches tall. The 4-inch pods grow in pairs. Resistant to powdery mildew and fusarium wilt.
- **Lincoln** Compact vines and tolerance for warmer weather. A good selection for smaller spaces in Texas gardens.

Green peas (Photo by Andrew Edmonson)

- **Little Marvel** (**American Wonder**) Heavy yields on 15- to 20-inch plants. Pods 3 to 4 inches long that contain 6 to 7 very sweet, dark-green peas per pod.
- **Mr. Big** A bush variety 2 to 3 feet tall that produces extra-sweet, very large, double pods 4½ inches long with nine or ten peas each.

Oregon Trail A bush variety that produces dark green, twin pods of very good quality. Resistant to mosaic virus, powdery mildew, and fusarium wilt.

Wando A heavy producer of 3- to 4-inch pods that contain seven to nine peas each. Somewhat tolerant of warmer weather.

Snow Pea Varieties

Carouby de Mausanne Purple flowers that produce large, tender pods on 3- to 5-foot vines.

De Grace An heirloom dwarf form grown in America since the early 1800s. Sweet, tender, medium-sized pods.

Mammoth Melting Sugar Sweet 4- to 5-inch pods produced on 4- to 5-foot vines. Sweet even after they mature.

Oregon Giant Sweet, wide pods 5½ inches long by 1 inch wide. Bush-type plants 30 to 36 inches tall. Resistant to mosaic virus, powdery mildew, and fusarium wilt (race 1).

Oregon Sugar Pod II Bush-type, heavy producer of 4-inch, sweet pods. Some mildew resistance.

Sugar Snap Pea Varieties

Cascadia Plump pods 3½ inches long. Resistant to mosaic virus.

Sugar Ann A very early variety that produces 3-inch pods on 2-foot vines.

Sugar Snap Very sweet, 3-inch pods on 6- to 8-foot vines. Tolerant of wilt.

Sugar Sprint Very sweet, stringless pods on 2-foot vines. Resistant to mosaic virus and powdery mildew.

SOUTHERN PEAS
Vigna unguiculata

Description

Southern peas, also known as black-eyed peas, crowder peas, field peas, and cow peas, are more closely related to beans than peas in cultural requirements. They need to be planted in warm weather and thrive in hot summers. These humble little vegetables are hard workers in Texas gardens and have very little disease or insect problems. Black-eyed peas can be eaten fresh, and the pods are edible if they are picked young. They can also be frozen or dried.

Southern peas

Varieties

Big Red Ripper Reddish-green, 10-inch pods, with up to eighteen peas per pod. Can be eaten fresh or dried. Very heat tolerant.

Brown Crowder Very productive plant 2 to 3 feet tall. Pods 7 to 9 inches long with seven to nine light brown peas per pod.

Calico Crowder Buff-colored, flavorful peas with maroon splotches when dried.

Colossus Prolific variety that produces large 7- to 9-inch pods. Buff and purple peas easily shelled from the pods.

Mississippi Silver Silver pods about 6½ inches long with occasional rose streaking. Large seeds easy to shell from the pods.

Peking Black Crowder A vigorous variety that produces an abundance of large, black peas.

Pinkeye Purple Hull An early, semi-bush variety that produces white peas with purple or pink eyes.

Washday Very productive variety that produces medium-sized peas on half-runner vines.

Whippoorwill Believed to have come to America on slave ships. Highly productive vines around 5 feet long that produce 7- to 8-inch green pods tinged with purple.

LETTUCE
Lactuca sativa

❋ **DESIGN USES:** Lettuce varieties offer some of the most vivid reds and bright green colors in the edible landscape. Deeply veined and variegated va-

'Merlot' and 'Winter Density' lettuces, with kale background

rieties provide foliage interest as well. Plant lettuces in formal patterns, as borders, or as pocket plantings in a mixed border. Lettuce plants are pretty in the border when combined with pansies and alyssum.

Description

Lettuce can add a beautiful container planting or colorful border to the winter garden. Varieties come in bright red and green colors with combinations of smooth, ruffled, and oakleaf shapes to the leaves. The loose-leaf varieties perform better in Texas gardens than the Bibb or butterhead types. Loose-leaf lettuce leaves can be harvested with the "cut and come again" method of harvesting the lower leaves and leaving the rest of the plant to produce, or the entire head can be harvested, removing the roots when the heads are pulled.

Cultivation

Lettuce requires a soil with a pH of 6.0 or greater, so some East Texas gardeners may need to adjust their soils with additions of lime. Plant lettuce in full sun or part shade in beds with plenty of well-decomposed organic matter worked in prior to planting. The tiny seeds can be sown directly in the bed, or they can be started in containers to extend the season. Lettuce does not germinate well in hot weather and can be damaged in freezing temperatures, so the window of opportunity for lettuce growing can be rather short. Plantings from container grown plants is the best method. Fall plantings have a better success rate than spring plantings in Texas. Lettuce varieties with tight heads generally require a longer growing season, and it is difficult to grow mature heads before 90-degree temperatures set in if seeds are sown in the spring. Average temperatures of 80°F will cause leaf lettuce varieties to go to seed, and the leaves become tough and bitter.

If seeds are sown directly in the bed, they should be planted about 10 weeks before the first expected freeze. Seed can be started in containers a couple of weeks earlier. Successive plantings can be made every 10 to 14 days to extend the harvest. If seeding directly in the garden, make sure the bed has been raked smooth and has a fine texture, with no clods or rocks. Planting beds should be wetted thoroughly before planting and the seeds firmed into the soil with the back of a hoe to facilitate better contact. Do not allow soil to dry out during germination. Water seedlings with a very light mist of water to keep them from being washed down into the bed. Leaf contact with soil can cause damping off and other fungal problems.

Lettuce should have an eventual spacing of 4 to 6 inches for leaf lettuce and 8 inches for butterhead and romaine. They can be started at a closer spacing, and the young plants can be harvested and eaten as more room is made for the remaining plants. Lettuce can survive freezing temperatures by pulling up a loose mulch such as straw or pine needles around the plants. The mulch should be pulled away from the plants when temperatures warm. Floating row covers can also be used to protect plants during freezing weather. Romaine and other more thick-leaved varieties handle freezing temperatures more readily than the more delicate-leaved varieties.

Lettuce should never be allowed to dry out completely, nor should the plants be allowed to stand in wet soil. They require about an inch of water a week. Soil should be loose and rich with organic matter, allowing water to drain. Both seeds and seedlings will benefit from periodic applications of fish emulsion and seaweed.

Loose-Leaf Varieties

Australian Yellow Sweet, large, frilled chartreuse leaves. Some heat tolerance.
Black Seeded Simpson Green leaves. Fast growing and tender.
Bronze Arrow Large oakleaf-shaped leaves with bronze-red tips.

Danyelle Deep red leaves on full, compact heads.

Deer Tongue Unique-looking, attractive lettuce. Leaves form a low rosette of broad, somewhat pointed leaves. Mild flavor. Good resistance to both warm and cold temperatures.

Drunken Woman Beautiful lettuce variety with ruffled, bright green leaves tipped in bronze-red. Mild, sweet leaves have crisp texture. Slow to bolt.

Flashy Butter Oak Green leaves with red mottling. Very showy with a mild, buttery flavor.

Lollo Bionda Lime-green sister of 'Lollo Rosso' adds a bright splash of color to the garden. Compact, crinkled, and flavorful leaves.

Lollo Rosso Selway Heavily frilled leaves with deep red margins.

Merlot Wavy, deep red leaves and open, upright heads. Good heat tolerance.

Mignonette Bronze Frilled bronze-green heads well adapted to warmer temperatures.

New Red Fire Large, red, sweet-flavored leaves. Good heat tolerance. Some resistance to mosaic virus.

Oakleaf Forms tight clusters of oakleaf-shaped leaves and has a very interesting appearance. Some bolt resistance.

Pablo Batavian Loose, ruffled, attractive green and red leaves.

Red Deer Tongue Triangular, tender, red leaves.

Red Sails Deep bronze-red color with a sweet flavor.

Loose-leaf lettuces, Natural Gardener, Austin, Texas

Red Velvet Dark red to purple-black, mild-flavored leaves that are green on the underside. Slow to bolt.

Royal Oakleaf Deeply lobed, bright green leaves.

Salad Bowl Deeply lobed, green leaves. Good heat resistance.

Simpson Elite A derivative of 'Black Seeded Simpson' with more heat tolerance.

Romaine Varieties

Cimarron Red Long, deep red leaves. Interior leaves creamy yellow.

Cosmo Green leaves 1 foot long and savoyed. Some heat tolerance.

Forellen-Schluss Green leaves with reddish

markings and firm with thick midribs. Good flavor. Bolts in hot weather.

Little Gem Leaves bright green with yellow hearts. Heads only 5 inches tall and nice for a border.

Outstanding Dark red leaves forming fairly tight heads. Some resistance to downy mildew and some heat tolerance.

Parris Island Classic romaine variety. Tall, 10- to 12-inch heads with ruffled leaves. Resistant to tip burn and mosaic virus. Some resistance to bolting.

Rouge d'Hiver Compact, 10- to 12-inch heads with red-bronze tips and green centers.

Nightshade Family

Description

Some plants in the Solanaceae family are actually poisonous, such as deadly nightshade and datura. The leaves, flowers, and stems of the edible plants of this family contain poisonous alkaloids. Even with these plants' darker side, a garden would seem incomplete without red ripe tomatoes, fresh potatoes, and rainbows of peppers and eggplants.

When designing a landscape with these plants, it is important to remember that some members of this family tend to look rather unkempt at times. While some plants in the edible landscape deserve a prominent place in the garden, some solanaceous plants are probably best tucked in behind other plants. Eggplant and pepper plants can be very pretty and ornamental (if they are not attacked by flea beetles). Tomato and potato plants

'Roma' tomatoes (Photo by Andrew Edmonson)

can be downright ugly at the end their production. The leaves of potato plants die to the ground before the tubers are ready to harvest. Tomato plants should be located behind other plants to hide their gangly, unkempt appearance as they mature. However, tomato plants are generally staked or trellised, and interesting "sculptures" might be created with different types of trellising.

Edible plants in the Solanaceae family have similar cultivation requirements, except potatoes, which are planted in the late summer and early spring in different areas of Texas. Because potatoes are unique in their requirements, their cultivation will be discussed separately from that of eggplants, peppers, tomatillos, and tomatoes.

Cultivation

Eggplants, peppers, tomatillos, and tomatoes are set in the ground in the late spring when soil temperatures have warmed to around 60°F. Set out tomato transplants around the last freeze date for the area and pepper and eggplant transplants about 10 days later. Start seeds 6 to 8 weeks prior to in-ground planting. There is not time for tomatoes, tomatillos, eggplants, or peppers to be sown as seed in the garden and produce flowers in time to set fruit in Texas. Once temperatures climb to the 90s, flower production drops off in most varieties and fruit will not set.

An additional fall crop can be expected from healthy pepper and eggplants when temperatures begin to cool. Tomatoes will probably need to be replanted around 100 days before the first anticipated killing frost for fall production. Generally, most varities of spring-planted tomatoes do not produce an additional fall crop.

For indoor seed planting, plant seeds in flats and place the flats in a sunny location. The seeds should not be allowed to dry out completely, but the soil should drain well to prevent "damping off" (a soilborne fungal disease) from killing the young seedlings. After the seedlings are 1 to 2 inches tall, they can be gently pried from the flats and transplanted into 4-inch pots or similar-sized

'Matt's Wild Cherry' tomato (Photo by Andrew Edmonson)

containers. If the seeds are started 8 weeks before time to set them in the garden, they may become large enough to be transplanted into gallon-sized containers before they are set in the garden. Early planting will result in larger transplants and earlier fruit production. The race for the earliest tomato and the need to set fruit set before temperatures climb into the 90s will cause gardeners to use all kinds of tricks to hasten the planting date.

The young plants need to be "hardened off" before they are planted in the beds. This is the process of slowly adapting the seedlings to the outdoor environment. Plants are brought outdoors each day for a couple of weeks before planting, and they are brought back indoors at night. If the temperatures are unseasonably high, the plants should be protected from strong outdoor sunlight in the afternoon. They should also be protected from strong winds. If there is not time or room to grow plants from seeds, young plants can be purchased at nurseries and garden centers. If a particular variety is desired, the selection may be somewhat limited.

Solanaceous plants should be planted in well-drained beds with plenty of organic matter. They are heavy feeders and will respond well to applications of well-decomposed compost. Beds should be worked 8 to 10 inches deep before planting. Additions of organic phosphorus will encourage good bloom set, which is essential to good fruit production. Sources of organic phosphorus are generally "slow release," meaning they will take some time to become available to the plant's roots. It is a good practice to add phosphorus at least a month or two before planting, about the time seedlings are started. Solanaceous vegetables prefer a pH between 5.5 and 7.5, and lime may need to be incorporated in some East Texas gardens.

It is also a good practice to water the seedlings with fish emulsion and seaweed and to water the plants in with this mixture when they are set in the garden. After they are in the ground, the plants will benefit from foliar applications of fish emulsion and seaweed every couple of weeks. Wait to apply mulch until the soil has warmed. Mulching at this time will give the soils time to warm com-

pletely, as well as help conserve moisture and keep weeds from competing for nutrients. Consistent watering is very important for good fruit production, especially for tomatoes. Blossom end rot on tomato fruit is commonly associated with inconsistent, uneven watering.

The planting method for tomato sets is unique. The other plants discussed in this section are planted at the same level at which they are growing in the container. The lower set of leaves on tomato sets can be removed and the plants set more deeply, covering the wounds where the leaves were sprouting. This is not required; however, if this planting method is used, the stems will produce additional roots to anchor the plants and make them stronger. Protect tender seedlings from strong winds and extreme temperatures.

EGGPLANT
Solanum melongena

❋ **DESIGN USES:** Eggplant leaves are large and bold textured, with various degrees of lobe depth. Eggplants can be a very dramatic summer planting. There are many varieties of eggplant, with various shades of purple, green, yellow, or white fruit that grow in a variety of round or elongated shapes. The pendulous fruit is very pretty. Plants should be spaced about 2 feet apart in garden beds, or they can be very attractive as container plantings.

Description
Eggplants are well suited for growing conditions in Texas and are more resistant to insect and

Eggplant fruit (Photo by Andrew Edmonson)

disease problems than other members of this family. Eggplants will continue to set fruit until the first killing frost, with slowing production in the summer months. Harvest the mature fruit often to encourage new fruit production. The fruit should be harvested before the skin becomes tough, or the pulp will tend to be bitter.

Varieties

Black Beauty A very popular and prolific variety introduced in 1902. Plant grows 2 to 3 feet tall and produces 1- to 3-pound oval fruits with shiny, purple skin.

Blush Not the 'Purple Blush' variety introduced by Burpee Seeds, which is a hybrid. Fruit creamy white with lavender streaking. Flesh has good flavor.

Goyo Kumbo A prolific producer of 3-inch, bright red-orange fruits. Striking 4-foot-tall plants when filled with brightly colored fruits.

Green Apple A small, round, light green eggplant with mild-flavored flesh.

Listada de Gandia Very pretty 5- to 6-inch, oval eggplant with purple streaking. Resembles colored eggs. Skin is thin and fruit does not need to be peeled. Pretty but tough plant. Tolerates heat and short dry periods.

Long Purple An Italian variety that produces 10-inch, narrow, purple fruit, usually about 1 inch in diameter.

Louisiana Long Green Fruit light green, narrow, about 8 inches long with a mild flavor.

Ping Tung A truly striking Taiwanese variety with long, slender, bright purple fruit.

Rosa Bianca A beautiful eggplant with plump, round, white fruit with purple streaking. An Italian heirloom with mild flavor.

Rosita A very good producer of bright purple to lavender-pink, pear-shaped fruit, generally 6 to 8 inches long and 3 to 6 inches across. Taste is mild and skin is not bitter.

Rotonda Bianca Sfumata di Rosa An Italian heirloom variety with large, white fruit tinged with shades of pink, lilac, or purple.

Thai Green Some varieties small and round with green stripes, and some long, slender, and light green. Commonly used in stir-fry and curry dishes in Thailand. Prolific plants that can withstand some dry conditions.

PEPPER
Capsicum annuum

❋ **DESIGN USES:** Pepper plants add color and interest to the garden. Looking like bulbs on a Christmas tree, varying fruit shades of red, yellow, and green can occur on an individual plant as the peppers reach different stages of maturity. The plants are usually attractive through the summer, with bright green leaves, and deserve a prominent place in the garden. Pepper plants are very attractive when planted as taller elements in containers with draping plants like thyme and trailing rosemary spilling over the sides.

Description
Peppers set most of their flowers when temperatures are below 90°F, so it is important to plant peppers in the beds with enough time to set fruit

Bell pepper (Photo by Andrew Edmonson)

before temperatures become too hot. Even though pepper plants will usually last through the hot summers and produce a second crop of fruit in the fall, additional fall plantings can be made around 100 days before the anticipated first killing frost. Hot peppers tend to take the heat and produce fruit in the summer more readily than sweet peppers.

Peppers have developed a natural defense to most animal predators. Most varieties have a degree of spiciness that is uncomfortable to the palates of most animals. Bell peppers are a variety of *C. annuum* that registers zero on the Scoville scale, which measures the spicy heat of peppers. Other varieties have varying degrees of heat from fairly mild to extreme heat.

Cultivation
Space plants about 18 inches apart in well-prepared beds with plenty of organic phosphorus to encourage blooming. Planting in the afternoon shade of tall screening plants will offer some protection to the fruit from sunscald. Pepper plants are usually shallow rooted, and it is not uncommon for them to fall over after a strong storm. If this happens, simply firm the soil around the roots and stake the plant if necessary. It is important to mulch pepper plants after the soil has warmed.

Pepper seeds are attached to the fruit wall with a white membrane. The seeds are easy to remove from the flesh and can be saved for future planting. If this is desired, the fruit should be left to mature completely on the plant before harvesting for seed.

Hot Varieties

Aci Sivri Large plants, 3 to 4 feet tall, and prolific producers of 7-inch, slender, curved peppers. Moderately hot peppers bright red when ripe.

Aji Amarillo Plants 3 feet tall that can produce fifty to one hundred peppers per plant. Fruits 2 to 3 inches long, mildly hot when they are green and ripen to a hotter yellow pepper.

Anaheim Chile (**Chile Verde**) Developed from the ancient pasilla chiles. Fruit 7½ by 2 inches. Often picked when green but ripens to a bright red. The variety commonly packed in small cans and used in Mexican dishes. Resistant to tobacco mosaic virus.

Ancho Large peppers often used for chiles rellenos, although other types can be used for this purpose. Have a little heat but relatively mild. Turn from dark green to a deep rust-red when mature.

Black Pearl Edible but commonly grown as an ornamental plant. Small, purple-black, very hot fruit matures to red. Grows in clusters on 2-foot plants that have deep purple stems and leaves. Plant very pretty in a container or as a border.

Cayenne Peppers about 6 by 1 inch with medium heat. Red when fully mature.

Habanero Little pepper that packs a punch. Mature fruit squat, about 1 to 2 inches long, and extremely hot. Matures from green to orange to bright red.

Hot Banana Pepper A hot version of the banana pepper and similar in appearance. Medium heat.

Jalapeño The popular spicy pepper that gives Mexican food its heat. Fruits 2 to 3 inches long ripen from green to red. Chipotle sauce is made from dried, smoked jalapeños.

Pepperoncini Very flavorful, 3-inch peppers that are often pickled. Greek (2 to 4 inches long) and Italian (about 2½ inches long) varieties.

Serrano Plants 30 to 36 inches tall that produce numerous 2-inch, dark green, very hot fruits.

Serrano Tampiqueño Plants 30 to 36 inches tall. Very popular in Mexico, probably because of the heavy yields of very hot peppers 2½ by ½ inches. Peppers increase in heat if left on the plant to turn from green to red.

Thai Hot Very hot, little red pepper. Low-growing plants (about 6 inches) will trail over a pot or make a pretty border.

Sweet Varieties

California Wonder A bell pepper can grow to 4½ inches and has good yield on 24- to 30-inch plants.

Cubanelle Produces 6-inch, tapered peppers that mature from green to red.

Doe Hill Golden Bell A four- to six-lobed, flattened, bell variety that ripens to a bright orange. Plants are 2 feet tall and good producers of fruit 1 by 2¼ inches.

Jimmy Nardello Long, slender fruits that are

sweet and very flavorful. A good variety for drying.

Marconi Good yields of 7-inch, curved peppers. Heat tolerant.

Melrose An early variety with a good yield of peppers 4 by 2 inches.

Miniature Bell Small bell peppers available in chocolate, red, and yellow varieties. Perfect for container growing.

Papri Sweet Plants about 3 feet tall that produce 6-inch paprika peppers. Allow to mature to a bright red before harvesting. A good variety for drying.

Sweet Banana Yellow-green, 6-inch, banana-shaped peppers that ripen to orange and then red if left on the plant.

Jalapeño pepper plant

Sweet Chocolate Very sweet pepper with dark brown skin and red flesh.

Sweet Red Cherry Relatively small, about 1 to 1½ inches in diameter.

POTATO
Solanum tuberosum

☀ **DESIGN USES:** Potato plants are not particularly attractive. The foliage will die back when it is time to harvest, making them even less appealing. Plant more attractive plants around potatoes to help disguise the plants' homeliness. Hills can be evenly spaced throughout the garden as architectural elements.

Description
Potato plants are usually grown from seed potatoes, which are small whole potatoes or larger ones cut in pieces. Seed potatoes can be ordered from a supply nursery, or they are usually available at local garden centers. Be sure to order from reputable suppliers. Inspect the tubers before purchasing for any disease or insect infestation. Tubers can be infected with late blight (*Phytophthora infestans*), a type of fungal disease, which can remain in the soil, on the tubers, or on plant leaves from season to season. Applications of sulfur have traditionally been used as an organic fungicide to treat cut tubers and as additional applications during the growing season. Potatoes are grown in cooler months when sulfur should not burn the plants, but sulfur can accumulate in the soil over time.

Potato harvest (Photo by James Beesley)

Cultivation

Cut seed potatoes into 2- to 3-ounce pieces a day or two before planting, making sure each piece has at least one "eye." Vegetative growth from the eyes produces new potato plants. Shake them in a paper bag with sulfur powder, which helps prevent fungal problems. Then allow the cut pieces to dry out or "cure" before planting. Plant the seed potato pieces 2½ to 3 inches deep, about 30 inches apart in raised beds or hills (6 to 10 inches high) about 4 weeks before the average last killing frost. It is a good idea to treat the cut tubers with an organic fungicide applied to the soil as a drench at planting. East Texas gardeners will be happy to know that potatoes like acidic soils. Alkaline soils can harbor potato scab.

Potatoes can also be grown from seeds planted 3 to 4 weeks before tubers, around the end of January. Potato plants produce a pulpy seedpod similar to a tomato from which the seeds are collected. To collect the seeds, shred the pod into a bowl of water to dislodge the seeds from the pulp. Let the seed mixture ferment overnight, allowing the seeds to separate and sink to the bottom of the bowl. They can then be dried and stored for future planting. Plants grown from seeds will produce only small tubers, or "tuberlets." The yield will not be as prolific as that from the cut tubers. These small tubers can be saved in a dry area for planting the following season. Do not store any diseased tubers, as they will infect the entire batch. Subsequent plantings from seeds should have increased yields because the resulting plants become better adapted to growing conditions in the area over time.

For the first several years, potato seed propagation is more a science experiment than crop production method. The plants grown from the seeds are not going to be the same as the variety from which they were harvested. Some will have desirable traits, and others will not. The plants should be culled, allowing only the most healthy and disease-free plants to mature and bloom. The

Potato seeds (Photo by Rebsie Fairholm)

seeds produced in this selective-breeding process are harvested and planted the following season, and the process repeated until the desired traits are produced.

Whichever propagation method is used, potatoes require deep, loose, well-drained soil for good production. Prepare the beds before planting with plenty of well-decomposed organic matter and an organic source of phosphorus. High-nitrogen soils will produce potato foliage rather than tubers so any high-nitrogen source should be avoided. Plant potatoes during the late winter months. Once the leaves are 6 inches above the soil line, lift the leaves and pile compost or dirt round the base of the plants. Repeat this process after the leaves have grown another 6 inches. This is referred to as "hilling." No additional fertilizer will be required during the growing season. Potato plants draw nutrients from soil amendments that have been added before planting. It is very important to provide consistent watering. Even watering, usually every 5 to 7 days, is necessary from planting time until harvest.

Potato plants will benefit from an application of 2 to 3 inches of loose mulch like straw or pine straw. The mulch can be pulled up over the leaves to prevent freeze damage in hard freezes and then pulled back from the plants when the temperatures warm. It is important to pull the mulch away from the plants to allow good light and air circulation.

Potatoes are ready to harvest when the tops turn yellow and die. "New" potato varieties can be harvested during the growing season, but russet or baking varieties are generally left until the tops have died. The potatoes can be harvested by digging outside the hills and coming up underneath the tubers rather than digging straight down, which could slice through the tubers. Potatoes can be stored in a cool, dry location after they are washed. Be careful to discard any diseased or insect-infested potatoes before storage. Potato eyes should be scrubbed from the tubers before they are eaten. The eyes of potatoes, as well as the leaves, are mildly poisonous.

Varieties

Some hybrid varieties are included in this listing, because potatoes are usually propagated vegetatively rather than by seeds.

Adirondack Red An early to midseason variety that produces fair yields of slightly flattened, purplish-red tubers with pink to red flesh. Flavor good for boiling, frying, mashing, and salads. Plants have a sprawling growth habit.

Caribe An early variety that has purple skin and white flesh. Medium-sized plants. A good-yielding variety that can be used for boiling, baking, frying, or dug early for new potatoes. Resistant to scab. Stores well.

Carola A yellow-fleshed variety with light yellow skin. Produces a good yield of medium-sized potatoes used for baking or frying.

Cranberry Red A pretty, early to midseason variety with red skin and lighter red flesh and good yields. Low in starch.

Dark Red Norland A dark red potato with white flesh that produces early. Highly resistant to scab and resistant to leaf rollers and some viruses.

French Fingerling A mid- to late-season variety on tall, spreading plants that produce large fingerling potatoes with rose-red skin and yellow flesh with pink coloring. Resistant to scab.

Kennebec A lightly russeted variety that produces large potatoes, commonly used for baking. Resistant to late blight, scab, and some viruses. Stores well.

Purple Viking A unique variety that produces large, round, purple-skinned potatoes with white flesh on compact plants. Some resistance to scab and leafhoppers.

Red Cloud A red variety with white flesh produced on medium-sized, spreading plants. Highly resistant to scab, interior rot. and early blight. Tolerant of heat and occasional dry conditions. Stores well.

Red Dale An early-maturing variety that produces large potatoes with red skins and white flesh. Can be planted in the fall. Good resistance to verticillium wilt.

Red Gold A midseason variety with red skin and waxy, yellow flesh. Some scab resistance. Does not store well for long periods.

Red Pontiac (**Dakota Chief**) An early-season variety that produces round, red tubers with white skin. Some drought tolerance and medium resistance to scab. Stores well.

Rose Finn Apple Fingerling A little fingerling potato with waxy, rose-colored skin and good yields on upright plants. Resistant to scab.

Rose Gold Red skin and creamy yellow flesh with good flavor and disease resistance.

Yukon Gold A commonly grown, yellow-fleshed variety with buff-colored skin. A very good producer that stores well.

TOMATILLO
Physalis ixocarpa

❋ **DESIGN USES:** Tomatillos grow from 2 to 5 feet in height and have smooth leaves. The fruit is similar to a hard green tomato and is yellow-green when it ripens. The fruits dangle from the stems like little paper lanterns. Give tomatillo plants a prominent place in the garden, as they are sure to be a conversation starter.

Description
Tomatillos are referred to as "husk tomatoes" because of the papery husks that surround the fruit. This plant has been cultivated in Central and

'Violet' tomatillo (Photo by Phil Christopher)

orous producer of fruit 1½ inches in diameter. Has a sharper taste than many varieties.

Everona Large Green Green fruits 2 inches in diameter.

Purple Coban Green fruit 1 inch in diameter with purple splotches.

Tomate Verde Plants generally 4 to 6 feet tall and may require staking or trellising. Fruit matures to a pale yellow color and grows to about 1½ inches in diameter. Flavor is a combination of sweet and sour.

TOMATO
Solanum lycopersicum

❈ **DESIGN USES:** It is a good thing that tomatoes taste good, because they become increasingly less attractive as they mature. Plant more attractive plants in front of tomato vines to hide the lower leaves and stems as they brown out and die. Tomato plants are often called vines. While they do not produce tendrils like many vining plants, the stems are usually not strong enough to hold themselves upright without staking. Tomatoes are susceptible to numerous fungal diseases, so letting them trail on the ground is not advisable. Trellising is not limited to tomato cages, and this is an opportunity to be really creative with various trellising techniques to add interest and whimsy to the garden. Tomato plants are generally spaced 3 to 4 feet apart in the garden.

South America for centuries. Tomatillos are used in fresh or prepared salsas, the most well known being salsa verde. Cultivation requirements are similar to those of tomatoes, and they are relatively disease and insect resistant. Tomatillos may require more than one plant to produce seeds, as they are highly self-incompatible, meaning they will not self-pollinate.

Varieties

Cisneros Grande High yields of large fruit 2½ inches in diameter. Fruits ripen to light green or yellow and become sweeter if left on the plant until the husks dry.

De Milpa A purple-fruited variety that is a vig-

Tomato plants (Photo by Andrew Edmonson)

Description
Tomatoes are highly prized by gardeners, and rightfully so. Nothing compares to a fresh ripe tomato from the garden, certainly not the year-round offerings from the grocery store, which are picked before they are completely ripe and bred for their ability to be transported rather than flavor. Tomatoes have been hybridized for many reasons in the effort to produce the perfect plant and fruit. In general, many hybrid cultivars offer disease resistance that heirloom varieties do not possess. Remember that hybrids will not come true from seeds if seed saving is desirable.

Trying to cover the vast number of named tomatoes that have been cultivated is beyond the scope of this book. Instead of attempting to create a comprehensive list of varieties and cultivars, a description of types and listing of a handful of varieties will have to suffice. The fun of selecting specific varieties will be left to the gardener's favorite winter ritual (drooling over the seed catalogues).

There are a few factors that will help narrow the field when selecting tomato varieties. Tomatoes are divided into "determinate" and "indeterminate" types. Determinate tomatoes are generally more compact and tend to produce all their fruit at one time. It is possible to plant early-, mid-, and late-season determinate tomatoes for continuous harvest. One thing to keep in mind when selecting determinate varieties for Texas gardens is that temperatures may rise too fast for late- or even midseason types to set flowers and produce fruit.

Many gardeners find they have more than enough tomatoes at once, and they might prefer indeterminate varieties, which tend to be more gangly and produce over the season. Theoretically, indeterminate tomato plants will last through the summer and produce a second crop in the early fall. In reality, the plants will probably succumb to disease, and a second planting in mid-August will have to be made for a fall crop. (Seeds should be started in mid-June to early July for August planting.)

Another consideration is size. There are cherry, plum, and regular-sized varieties. Cherry tomatoes are bite-sized and are often used for salads or relish trays. Plum tomatoes are larger than cherry tomatoes and more elongated. Plum tomatoes have multiple uses, including canning or processing for sauces. The larger, sandwich-sized tomatoes are called "slicers."

Disease resistance is another important consideration when selecting varieties. Tomatoes are sometimes labeled with one or more of the letters *V*, *F*, or *N*, relating to verticillium, fusarium, and nematode resistance. This is the trifecta of tomato disease resistance.

Varieties

Cherokee Purple Indeterminate heirloom (80 days). Large, dark-skinned tomatoes. Very sweet fruit with red flesh and green pulp around the seeds, making it a very interesting and flavorful slicer. Tolerates hot temperatures.

Eva Purple Ball Indeterminate (70 to 75 days). An early producer of medium-sized, pinkish-purple tomatoes with some resistance to cracking. Tolerates hot, humid weather. Some disease resistance.

Green Zebra Indeterminate heirloom (77 days). Medium-sized green fruit with dark green, longitudinal stripes and bright green pulp.

Illini Star Indeterminate (69 to 80 days). Medium-sized red fruit.

Marglobe Semi-determinate (75 days). A cross between 'Marvel' and 'Globe' varieties. Medium-sized red, globe-shaped fruits. Resistant to verticillium and fusarium wilts. 'Marglobe Supreme' (also semi-determinate) matures earlier at 69 to 80 days.

Matt's Wild Cherry Indeterminate (60 days). Early-season variety that produces clusters of deep red fruits ¾ inch in diameter. Can be grown without staking. Grows wild in eastern Mexico. Resistant to late blight.

Principe Borghese Determinate (75 days). An Italian heirloom variety commonly used for drying. A flavorful variety that produces prolific numbers of small red tomato clusters.

Roma VF Determinate (75 to 80 days). Produces pear-shaped tomatoes 2½ to 3 inches in diameter that can be eaten fresh. Good for processing as canned tomatoes or as tomato paste. Resistant to alternaria and verticillium and fusarium 1 wilts.

Thessaloniki Indeterminate (66 to 72 days). A Greek heirloom variety that is a heavy producer of 5- to 7-ounce red fruits. Resistant to sunscald and cracking. Stores well.

Okra
Abelmoschus esculentus

☀ **DESIGN USES:** Okra is a relative of the ornamentals hibiscus and turk's cap, and the beautiful pale yellow or pink flowers are equally showy. The horn-shaped okra pods perched on top of the stalks make a dramatic showing as well. Some varieties have red stems or leaves that make a beautiful backdrop for yellows, oranges, silvers, and purples. Okra plants can grow 8 to 10 feet in height, so it is best to place them toward the back of the bed where they will not shade or hide shorter plants. They also grow well as container plants.

Okra flower

Description

After many vegetables begin to look a little worn in the heat of the Texas summer, okra plants stand proud and beautiful in the garden, with attractive yellow or pink, hibiscus-like flowers, and they also produce a very tasty vegetable. Even though many people do not like okra because of its "slimy texture," this edible plant is so care-free and beautiful it could well become a favorite in the summer garden, even if it is just grown for the showy, hibiscus-like flowers. Okra does not usually have many problems with insects and disease. If the soil is healthy and the beds are kept clean of weeds and debris, the plants are usually healthy. Root-knot nematodes and fusarium wilt may already exist in the soil, which can cause problems.

Cultivation

Okra seeds are usually sown directly into garden beds after all danger of frost has passed. The seeds can be soaked overnight before planting to hasten germination. It is also helpful to nick the seeds with a file or sandpaper to allow water to penetrate the seed coat. Although it is not essential for germination, the seed coat will allow the water in more quickly if this is done.

Okra plants have deep root systems and are able to reach deeply for nutrients, so beds will not need large additions of nutrient-rich compost. However, these plants do require well-drained soils. Lighten heavy soils with plenty of organic matter or well-decomposed compost. The plants do best in a soil pH above 6.0.

Okra does not benefit from early seeding. The plants can be stunted if they are seeded before nighttime temperatures are averaging below 50°F. Plant seeds in pre-moistened beds about ¾ to 1 inch deep, spaced 3 to 4 inches apart. When they are a few inches tall, thin the plants to 6 to 12 inches apart, selecting the healthiest seedlings.

Okra does not require an abundance of water but will produce higher yields with consistent watering. Do not allow the soil to become completely dry, which can cause tough, misshapen pods. Okra will need to be watered at least every 7 to 10 days during dry weather. The plants will benefit from an application of mulch when they are a few inches tall, about the same time they are thinned. Fertilize the plants with fish emulsion and seaweed every couple of weeks, and side-dress with compost every 3 to 4 weeks to encourage new growth.

Although okra will continue to grow and produce through the summer, additional plantings can be done through June to lengthen the season, and an August planting will produce a fall crop before freezing temperatures arrive. Fall planting should cease 90 days before temperatures are expected to be consistently below 70°F. The plants can be cut back to a few inches above the ground if production wanes, and a second flush of growth is likely to produce more pods.

Okra produces hibiscus-like flowers about 60 to 70 days after planting. The pods will be ready to harvest 3 to 4 days later when they are about 3 to 5 inches long. It is important to harvest the pods before they become too tough to be eaten, unless they are being grown for dried arrangements, which is a use for pods that are past their prime. Frequent harvesting will encourage new pod formation. Many okra varieties have small "spines" that can cause itching. Some varieties are spineless, like 'Clemson Spineless.' (This is not a comment on their character. It just means they do not have the irritating little hairs or spines that can cause itching.) If a variety is planted that has spines, it might be a good idea to wear long sleeves and gloves when tending or harvesting. It is also wise to take a pair of shears to clip the pods of most varieties, because they do not always pull free from the plants easily.

Varieties

Alabama Red An heirloom variety from Alabama that grows 5 to 7 feet tall and produces fat, red-green pods.

Beck's Big Buck An heirloom variety from Malcolm Beck, the famous plantsman and compost aficionado from Texas. Plants about 5 feet tall and prolific producers of fat, tender pods with heavy ribbing.

Bowling Red Red-stemmed variety that grows 7 to 8 feet tall with long, slender pods.

Burgundy Early-bearing variety with rich maroon-colored stems, leaves, and pods, making a striking conversation piece in the garden. Plants about 4 feet tall that produce spineless pods 7½ inches long.

Burmese Large, dramatic leaves. Pods 9 to 12 inches long, slightly curved, and almost spineless. Produced early and continue producing until frost. Pods light green when young and turn yellow-green as they mature. Tender pods somewhat less slimy than other varieties.

Clemson Spineless Plants grow 3 to 5 feet tall and produce straight, spineless pods.

Choppee Named for the Choppee Indians. A high-yielding, semi-dwarf variety that grows about 3½ feet tall. Tender, almost spineless pods.

Cow Horn An heirloom variety that grows 8 feet tall and produces heavy yields of 8- to 15-inch pods. More tender if picked when 5 to 6 inches long.

Eagle Pass A variety from Eagle Pass, Texas, that produces tender, large pods that are somewhat less slimy than those of other varieties.

Emerald An early variety that produces large, thick-walled, deep-green pods, which remain tender even when large. Plants grow to about 4 feet tall.

Hill Country Heirloom Red Named for the Texas Hill Country. Produces an abundance of green, tender pods with reddish tips and ribs. Plant stems also red. A strikingly beautiful addition to the edible garden.

Jade An early-maturing, prolific variety that produces dark green pods that remain tender up to 6 inches long. Plants grow to about 4½ feet tall.

Jimmy T's Grows to about 5 feet tall and produces many pods if kept picked. Pods 8 to 12 inches long, but flavor best if harvested around 4 inches long. Some pods have spines, and others smooth.

Perkins Mammoth Long Pod Grows to about 5 feet tall and produces 6- to 9-inch, thick, tapered pods.

Philippine Lady Finger Grows up to 10 feet tall and produces long, smooth pods. Requires a long season to mature.

Star of David A very productive variety from Israel that grows 8 to 10 feet tall and produces 5- to 9-inch pods. Flavor is strong and may not be mild enough for all but true okra fans. Harvest pods when young and tender. Tolerant of root-knot nematodes.

Stewart's Zeebest A spineless variety that produces heavy yields of very tender, unribbed pods.

Vidrine's Midget Cowhorn A dwarf variety from Louisiana. Plants about 3 feet tall but produce white-green pods up to 15 inches long.

Onions

Description

Edible plants in the Allioideae family are called alliums, and they include garlic, leeks, onions, and shallots. There are some unique characteristics to each of these plants, but they all form edible bulbs that require well-prepared beds. Onion and shallot leaves are round and hollow,

Green onions

and garlic and leeks have flat, strap-type leaves. The leaves and flower heads grow from the base of the plant and are referred to as the "scape." Alliums are relatively care-free. The strong scent these plants emit can deter many garden pests, and they are good to plant throughout the garden for this reason.

Cultivation

Bed preparation is very important for alliums. They should be planted in full sun and in raised beds with plenty of organic matter and organic phosphorus worked into the top 6 inches of the soil. Loose, rich soil allows the bulb to mature freely, without rocks or heavy soil encumbering the growth. Because alliums do not compete well with other plants for nutrients, all weeds should be removed before planting, as it will be difficult to weed around the bulbs without harming them.

GARLIC
Allium sativum

❋ **DESIGN USES:** Plant garlic as a frame or toward the back of the garden bed where the flower heads can be seen but the gangly leaves will not detract from other planting. The silver-green foliage and white flower clusters are really very pretty as a backdrop and are particularly well suited to English cottage and heirloom garden styles.

Description

Garlic is very care-free to grow and has few disease or pest problems. It is planted in the fall in most

Garlic drying at Green Gate Farms, Austin, Texas

areas of Texas, although garlic can be planted in the late winter to early spring in northern counties. The newly planted bulbs require a 1- to 2-month vernalization period of temperatures between 32°F and 50°F to initiate bulb growth.

Cultivation

Separate the bulb into individual cloves and plant in well-drained garden soil 2 inches below the ground and 4 to 6 inches apart. Plant with the pointed tip up and the root side down. Garlic, like onion sets, will benefit from an overnight soaking in water with fish emulsion and seaweed and a teaspoon of baking soda before being planted. The larger bulbs generally produce larger garlic. Shoots will emerge through the mulch in 4 to 6 weeks. Garlic can take some afternoon shade but will be more robust in full sun.

Garlic tends to go through a type of dormancy in the winter, when it will stop growing. It will start growing again in the spring and be ready to harvest when the tops die back in the late spring and early summer. Be sure to remember where the plants are. They can be a little tricky to locate after the tops are gone. If a wayward bulb is left in the ground, it will just continue to grow into the next season. Be sure to save some of the cloves for the coming fall planting.

Varieties

There are two types of garlic, hardneck and soft-neck. Certain varieties of both types will do well in Texas gardens. According to Bob Anderson of Gourmet Garlic Gardens in Bangs, the hardnecks that grow well in Texas are the early-season Asiatic and Turban varieties and the late-season Creole varieties that store better. Of the softnecks, Bob recommends the silverskin and artichoke varieties. The silverskins are more temperamental about hot temperatures; however, they store very well and are sometimes risked for this reason. The following is a brief list of varieties.

Asian Tempest Strong flavor. Asiatic variety.
California Early Prolific and, with 'California Late,' probably the most common type grown in the United States. Mild flavor and very good storage characteristics. Artichoke variety.
California Late Strong flavor and good heat tolerance. A very good grower with twelve to sixteen cloves per bulb. Stores well. Artichoke variety.

Creole Medium sized with a full flavor and eight to twelve cloves per head. Can tolerate warmer temperatures. Bulb skin white, but individual cloves have rose- to purple-colored skins.
Elephant Actually a leek, but looks and per-forms like garlic. A very large bulb up to 1 pound each. Easy to peel with a mild flavor. If left in the ground over the summer, can be a good indicator of when to plant garlic: when the green shoots start to poke their heads out of the soil in the fall.

LEEK
Allium ampeloprasum var. *porrum*

❋ **DESIGN USES:** Leeks produce very dramatic gray-green, strappy leaves, making a spectacular backdrop or frame for more delicately textured plants. The foliage color is very pretty when contrasted with warmer colors, especially bright yellows and oranges or mixed with purple hues for a cooler effect.

Description

Leeks grow to about 18 inches in height with flat, gray-green leaves. The 2-inch-diameter bulbs have a mild onion flavor and are commonly used for soups and sauces in French cooking.

Cultivation

Leeks need to be planted in full sun in loose, well-drained soil with plenty of organic matter. Sow seeds in the late summer for winter or spring

Leek plants

harvest, or they can be started indoors 3 to 4 weeks before the last killing frost and harvested in late spring before temperatures become too hot. The seeds should be planted ½ inch deep, 1 to 2 inches apart. They can be thinned as needed for cooking to allow the remaining plants more space. The edible part of the leek is the blanched leaves. The leaves are blanched white by layering 5 to 6 inches of mulch or other dry organic matter around the stems as the stalk grows. Harvest leeks with a garden fork or shovel by loosening the soil around the roots.

Varieties

American Flag Matures in 130 days. A large leek with a mild onion flavor.

Dawn Giant Matures in 98 days. A very large leek, reaching 15 inches in height and 2 inches across.

Giant Musselburg Matures in 100 days. Heirloom variety with a mild flavor.

King Richard Matures in 80 days. A long, slender leek that can be harvested early as a "baby leek." Mild flavor.

Lincoln Matures in 75 to 100 days. A large leek with pale green, relaxed foliage. Can be harvested at 50 days for "baby leeks." Mild flavor.

ONION
Allium cepa

❋ **DESIGN USES:** The edible portion of the onion is a bulb that grows underground, so the foliage will be the part of the plant that is seen in the garden. An onion's foliage can be a nice contrast when planted with more delicately textured plants. Onions are also considered a good companion plant in the belief that strongly scented plants deter unwanted pests in the garden. Plant among other edibles and flowers for a bold texture.

Description

Onions are bulbing plants that produce their leaves first and then their bulbs. Each leaf produces a ring of onion, so a larger clump of leaves will produce a larger bulb. The bulbs do not start forming until the day length is correct. Onions are categorized by their preference for long, intermediate, or short days, which relates to the amount of daylight hours required for bulb formation. In Texas, short-day-length varieties are the best choice to give the bulbs time to mature.

Cultivation

In the southern half of Texas, onions can be grown from seeds, which can be sown directly in the garden beds in mid- to late October and up to early December. The earlier the seeds are planted, the more time the leaves will have to grow, and later seed plantings may be damaged by hard freezes. The overall goal is to grow the plants to ¼-inch diameter before the first dormancy-inducing temperatures occur. Because of danger from freezes, seeding onions in the fall is not recommended for the northern areas of Texas.

Many Texas gardeners prefer to plant onion transplants. These little onion plants come in bundles (usually fifty to one hundred sets per bundle.) Onion transplants should be about the same diameter as a pencil or a little smaller when planted. Larger-diameter plants will usually go to seed, or "bolt," before they mature. These little onions should be planted ¾ to 1 inch deep in well-prepared garden soil 4 to 6 weeks before the last expected freeze date. Transplants will usually be somewhat dry when they come from the nursery. They can be soaked overnight in a fish emulsion and seaweed solution to revive them before planting. A teaspoon of baking soda can be added to the solution to help the sets fend off soilborne fungal pathogens.

Onions grown from both seeds and transplants can bolt before the bulbs have matured. Onions are biennials, meaning they flower and seed in the second year. In Texas, the weather can be 90°F one day and below freezing the next. This can cause the onion to be fooled into thinking it has come

Onions drying at Itz family farm, Fredericksburg, Texas (Photo by James Beesley)

through two full years in one season. If this occurs, the onion will bloom and the energy spent on bloom production will diminish bulb growth. In addition, flower stalks grow from the center of the bulb, and the green growth can cause the onion to decay from the inside out after it is harvested.

Transplants should be planted in well-prepared beds. Soil pH below 6.0 is not a suitable growing medium for onions. Additions of lime may be helpful for low-pH soils. Plant onion transplants 4 inches apart for full-sized onions, or they can be planted 2 inches apart if the small onions will be harvested before they are mature. These are the popular "green onions" or "scallions" used in salads and sautés. These small onions are usually ready for harvest in 8 to 10 weeks after seeding.

Onions should be top-dressed with well-decomposed compost 2 to 3 weeks after the tops start to grow. In order to prevent fungal problems, do not let the compost touch the leaves. Mulch the onion plants with loose mulch, such as pine straw, at this time. Onions need about an inch of water a week. Do not allow onions to sit in soggy soils or become dry.

When the tops wither, harvest onions by lifting them out of the ground with a pitchfork or shovel. If no rain is expected, the bulbs can be laid on top of the ground to dry out for a few days before they are stored. They should be stored in a dry area, and it is best if the bulbs do not touch each other. Some people recommend drying the onions in long tubes, using netting or pantyhose with a knot or twist tie separating the bulbs.

Varieties

1015Y Texas Supersweet A large, yellow onion with good disease resistance that stores better than many short-day varieties.

Red Burgundy Red skin with red- and white-ringed flesh, 3 to 4 inches in diameter. Mild flavor.

Texas Early Grano 502 Straw-colored skin with white flesh. Mild to medium pungency. Some resistance to pink root rot.

White Bermuda A longtime favorite used for green as well as full-sized onions. Does not store well.

White Granex Mild white onion. Does not store well.

Yellow Granex (**Vidalia**) A very sweet onion that has a flat bulb. Does not store well.

SHALLOT
Allium ascalonicum

✱ **DESIGN USES:** As shallots mature, the bulbs can lift above the top of the soil and splay out in a circle. They have a shorter growth habit, so shallots can be planted toward the middle of the bed or even as a part of a mixed border.

Description

Shallots are a mild form of onion used extensively in French, Indian, and Southeast Asian cooking. The cloves are planted in the fall in the southern half of Texas and in the early spring in the north-

Shallot plants (Photo by Victor M. Vicente Selvas)

ern half. Shallots, like other allium members, respond to day length for their growth.

Cultivation

Plant the shallot bulbs 6 inches apart in well-prepared garden soil at a depth of 1 inch so the tip is even with the soil surface. Plant the bulb with the root side down, and firm the bulbs in the soil by pressing fingertips around the outside. Water the sets in with fish emulsion and seaweed. Be careful that fresh applications of compost and mulch do not come in contact with the tips, as this might cause a fungal problem. Do not allow the soil to dry out, but it should not be soggy either. Top-dress the plants with well-decomposed compost in the spring when new growth emerges. Shallots are harvested in the early summer when the tops die down. They can be stored after drying. Some of the bulbs may be kept for fall planting.

Varieties

Dutch Yellow Large, gold-skinned shallot that stores well.

French Red Reddish-pink bulb with purple-pink flesh. Mature bulbs 1 to 2 inches in diameter. Widely adapted with a good flavor but does not keep as well as other varieties.

Gray Griselle A different species (*A. oschaninii*). Hard gray skin and pink-tinged flesh. Bulbs 1 to 1½ inches in diameter. Flavor very rich and earthy.

SWEET POTATO
Ipomoea batatas

✸ **DESIGN USES:** Sweet potatoes are tubers, growing below the ground. The foliage is a cheerful display of masses of heart-shaped leaves that blanket the ground, providing beautiful green contrast to other plants. They should be planted in a mass as a single planting in the foreground of the bed with taller plantings behind.

Description

Sweet potato plants love hot weather and are a very good selection for summer gardens in Texas. Edible sweet potatoes are different from the ornamental sweet potato vines commonly found in nurseries, which are grown for their

Sweet potato plants

colorful foliage. The tubers come in a variety of colors, including red, orange, purple, brown, and white. The flesh can be white, yellow, orange, or purple.

Sweet potatoes are sometimes referred to as "yams," which are similar in appearance but from a different family. Yams are in the Dioscoreaceae family, while sweet potatoes are in the Convolvulaceae, or morning glory family, and have a similar showy, bell-shaped flower that is white with a dark pink throat. Flowers are sensitive to day length and rarely bloom if days are longer than 11 hours, so bloom is scarce in Texas. The leaves are broad and smooth with varying degrees of depth to their lobes.

These delicious and highly nutritious tubers are a staple food crop for many countries and are especially important to hotter regions of the world where other crops might not survive the harsh climates. Sweet potatoes have been grown in Texas since the early 1900s and were the predominant vegetable crop for the state in the 1920s and 1930s.

Cultivation

Sweet potatoes are grown from "slips," adventitious roots that grow from the tuberous roots. Many grade-school students will remember growing these slips in science class, by placing the root of the sweet potato in a jar of water until little sprouts grow from the tubers. Cultivation is similar if gardeners want to grow their own slips. Start with disease-free tubers and lay them on their side in a bed of sand in a cold frame where temperatures can be maintained at 80°F. Cover the roots with about 2 inches of sand. The sand must be moist and not allowed to dry out. After the little slips begin to show above the sand, more sand may be layered over them to produce stronger roots. When the plants are about 6 inches tall, pull gently to remove them for transplanting into their permanent location. Do not plant the slips in open-air beds until all danger of frost is passed and soil temperatures are at least 60°F. Cooler temperatures can stunt plant growth. Space the plants about 12 inches apart.

Sweet potatoes prefer deep, sandy loam soil with a clay subsoil. Certain areas of Texas will be able to provide this condition more readily than others, especially East Texas, where most of the commercial growing is done; however, sweet potatoes should grow in most well-prepared beds. Since the desirable part of the plant is the root, it is important that the bed contains at least 8 inches of loose soil. A foot of loose, well-drained soil would be better. Additional organic matter is appreciated, but make sure it is not too rich, as this will encourage leaf growth rather than root growth. Sweet potato plants do not require a high amount of nitrogen. Foliar sprays of fish emulsion and seaweed are beneficial. It is very important that sweet potatoes are planted in well-draining soil, as they can be afflicted with a number of fungal diseases that are encouraged by soggy conditions. It is also important to rotate sweet potato locations in the garden and to buy slips from a reputable source to prevent common insect and disease problems.

Sweet potatoes are an especially good crop for Texas because they do not need an abundance of water and require 100 days or more of warm to hot weather to mature. Sweet potato tops do not die back when they are ready for harvest, as tops of Irish potatoes do. The only way to determine if the sweet potatoes are ready for harvest is to gently dig the soil around the plant and pull some tubers up for inspection. Be careful to note the date when the plants are set in the garden and begin inspection around the "days to maturity" date for that variety. To avoid fungal diseases, dig when the soil is on the dry side.

Do not wash the sweet potato roots after harvest. This could rub off some of the skin and open the flesh to insect or disease infestation. Let them dry 3 to 4 hours out of direct sunlight. After they are dried, they can be "cured" by placing them in a loose medium such as shredded newspaper for a couple of weeks. After that, they can be stored. Do not allow the sweet potato tubers to become completely dried out, and be sure the temperature where they are stored does not fall below 55°F. They should keep for about 3 months if treated in this manner.

Varieties

Beauregard Large, uniformly shaped tubers with red-orange skin and orange flesh. Good yield and stores well. Resistance to soil rot and white grub but is susceptible to nematode damage.

Betty's Heirloom variety. Semi-bush type with pink-red skin and light orange flesh. Early producer.

Brinkley White Heirloom variety. Vigorous vines that produce large yields of cream-colored tubers.

Bush Puerto Rico Bush type with a more compact form than vining types. Flavorful tubers with yellow-orange skin and flesh.

Centennial Semi-bush type with copper-red skin and pale orange flesh. Early producer better adapted to heavy soils. Good drainage important to prevent fungal problems.

Edna Evans An heirloom, late variety with orange skin and flesh.

Excel Light copper skin and orange flesh. Early variety with good insect and disease resistance.

Frazier White An heirloom, early variety that produces white tubers with pale green leaves.

Ginseng Red Heirloom variety. Semi-bush type that produces large tubers with pink skin and orange flesh.

Golden Slipper Heirloom variety. Long slender tubers with light orange flesh and skin.

Hayman Heirloom variety. Vigorous vines that produce cream-colored tubers with flavorful creamy flesh. Tends to produce a large number of small tubers.

Hernandez Vigorous vines that produce good yields with dark orange flesh and skin. Resistance to fusarium wilt, southern root-knot nematode, soil rot, and internal cork.

Indiana Gold An heirloom, early variety with gold skin and orange flesh.

Ivis White Cream An heirloom, early variety that produces good yields of cream-colored tubers with light green leaves.

Jewel Good yield of squat tubers with copper-colored skin and bright orange flesh. Resistant to fusarium wilt, southern root-knot nematode, internal cork, and sweet potato beetle.

Korean Purple Heirloom variety. Purple skin and creamy white flesh. Yields and flavor very good.

Nancy Hall An heirloom, late variety that produces small tubers with creamy yellow skin and flesh. Very good flavor.

Pumpkin Yam An heirloom, early variety with good yields of tubers with orange skin and light orange flesh

Red Wine Velvet An heirloom, midseason variety with deep red skin and orange flesh.

Regal Purple-red skin and orange flesh. Good insect and disease resistance. Very good yield potential.

Ringley's Porto Rico An heirloom, early variety with ivy-shaped leaves. Tubers with cream-colored skin and light orange flesh.

Sharp An heirloom, midseason variety. Bush type with purple-tinted stems. Good yields of orange tubers.

Spanish Red Heirloom variety. Large semi-bush type that produces long, skinny tubers with red skin and white flesh. Very late producer.

Stevenson's An heirloom, midseason variety. Bush type with good yields of light orange tubers.

Sumor An heirloom, early variety with pale green leaves. White tubers with white flesh. More likely to flower than other varieties.

Vardaman Bush-type foliage is purple. Light orange tubers with orange flesh. Ninety days to maturity.

Violetta An heirloom, early variety that produces good yields of tubers with bright purple skin and white flesh. Good flavor.

Wakenda An heirloom, early variety with good yields of dark pink tubers with orange flesh.

White Yam An heirloom, midseason variety with above-average yields of white tubers.

Willowleaf An heirloom, early variety. Large semi-bush type that produces light orange tubers with orange flesh.

Saving Seeds

Saving seeds to replant the following season can be a very rewarding endeavor. There are a few things to keep in mind for successful seed saving.

1. Make sure the variety is open-pollinated rather than hybrid. Hybrid seeds cannot be relied upon to "come true." That is, the seeds may have the predominant characteristics of either parent but may not produce the same combination in future seasons.

2. It is important to know if the plant is an annual or a biennial. Biennials take two seasons to produce viable seeds. Swiss chard, beets, carrots, and most alliums and brassicas are biennials. It is common that the dramatic temperature swings in Texas will fool the plant into believing it has actually gone through two periods of vernalization (winter temperatures) and cause it to produce seeds in its first season. These seeds may not be viable.

3. There are two types of seeds, those that are dry and those enclosed in pulp. Beans and peas are examples of dry seeds and are commonly left on the vine until the pods are dried. Pulp-enclosed fruits like tomatoes and cucumbers will be left to mature for a little while but will need to be harvested be-fore they begin to rot, as this could infect the seeds. Winter squash is an exception. These gourds can be left on the vine for a couple of months after they mature to allow the seeds to fully develop.

4. Generally, fruit that is grown for seeds will be allowed to stay on the plant past the time of maturity to give the seeds a sufficient time to develop. This is especially true of dry-seed plants. It is important to keep the plants watered during seed formation to be certain the seeds are still receiving nutrients. Drip irrigation is a vastly superior method of watering to make sure fungal diseases are not spread to the fruit or seedpods.

5. Plants that have the same botanical (bino-mial) name will cross-pollinate with one another. Carrots can cross-pollinate with a wild carrot called Queen Anne's lace. Chard and beets will cross-pollinate with each other. Cruciferous vegetables, like cabbage, kale, and broccoli, are all the same species (*Brassica oleracea*) and will cross with each other. Even if they are separated by hundreds of feet, they can still share pollen through insect and wind pollination. Sepa-rating these plants by such a large distance

will probably not be practical in the small family garden. If seed purity is important, then only one species should be planted at a time, unless these distances can be achieved.

6. The problem with strongly outbreeding plants (plants that cross-pollinate readily) is that they can decline over time because of introduced genetic variability. This will be expressed in poor germination and yield, as well as a tendency to be more susceptible to disease and environmental factors. Strongly inbreeding plants do not decline over time, because it is their nature to reproduce with their own genetic information.

7. Edible plants have different methods of pollination. Some are inbreeding, and their flowers are enclosed during pollination, which does not allow other pollen to enter. These plants include most beans and peas, tomatoes, and peppers. (Runner beans and black-eyed peas cross more freely.) Other edible plants are outbreeding, including beets, chard, carrots, cucumbers, melons, eggplant, onions, leeks, pumpkin, squash, watermelon, and all of the brassicas. As a very rough rule, the Leguminosae family is inbreeding, and the Allioideae, Apiaceae, Asteraceae, Brassicaceae, Caryophyllaceae, Cucurbitaceae, and Poaceae families are outbreeding. Other families include both inbreeding and outbreeding plants. For example, hot peppers are more outbreeding than mild peppers. Likewise, tomatoes are mostly inbreeding, and eggplant is primarily outbreeding, even though they are in the same family. Corn can be inbreeding, but the male and female flowers mature at different times, so they usually pollinate with neighboring corn that has matured. This can become very confusing, but it is important to remember that plants must be in the same genus and species in order to cross-pollinate.

8. Larger populations of a species will help ensure genetic purity. Plants grown from seeds can become much more adapted to a site's growing conditions due to their genetic diversity. If seeds are gathered from only the healthiest plants, then the progeny will most likely be less susceptible to local diseases and environmental factors in future generations. It is important to remove poorly performing plants before they flower to promote the healthier genetic traits.

9. Monoecious plants have separate male and female flowers on the same plant. Dioecious plants have male and female flowers located on separate plants. Both the male and female flower parts are in the same (perfect) flower of bisexual plants. In unisexual plants, the (imperfect) flower contains only the female (pistil) or the male (stamen) reproductive parts. The following diagram shows the parts of a perfect flower. This diagram can be useful those who would like to "aid" the pollination process. The pollen-containing stamens can be removed to dab onto pistil stigmas to increase pollination, as well as to help ensure seed purity.

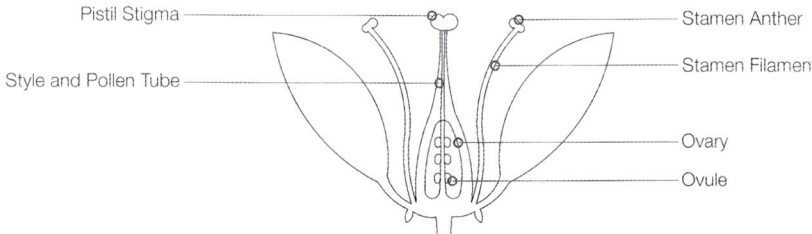

Pistil Stigma

Style and Pollen Tube

Stamen Anther

Stamen Filament

Ovary

Ovule

Perfect flower

Seed Harvesting

There are two types of seed-harvesting methods, one for the seeds enclosed in fleshy fruits, like tomatoes and melons, and the other for seeds in dry pods, like beans and okra.

FLESHY FRUIT

To separate seeds from fleshy fruits, first soak the pulp for a few hours and then rinse the seeds. Fermentation is another method in which the seeds are placed in just enough water to cover them and left to soak for a couple of days. They should be stirred twice a day. A white, frothy mold will form on the top, but this is not harmful to the seeds. After they have fermented, wash the pulp from the seeds. This process of fermentation helps kill disease and allows the pulp to separate more readily. Spread the seeds that have been separated from pulp on a cookie sheet, and stir often to make sure they dry thoroughly. Placing a sheet of parchment paper under the seeds will keep the seeds from sticking. Seeds need to be very dry before they are stored in an airtight container out of sunlight at temperatures below 95°F.

DRY SEEDS AND PODS

For seeds that are not in pulp, it is best to allow the seed heads and pods to dry on the plant. This will require a longer growing season, and seasonal garden planning should allow for the additional use of space. The seeds are usually ready to harvest when they come away easily when they are rubbed in the palm of the hand. They will usually be tan, brown, or black. Sometimes it is not possible to wait until the seeds are fully mature before they need to be harvested. Threats of rain, as well as rodent or bird foraging, can force an early harvest. In this case, the seeds can be brought inside to complete maturation. Hang the seedpods or seed heads upside down over a paper bag for a few weeks after they have been harvested. The seeds should be dried in a warm, dry location out of the sun. If the seeds shatter easily from the pods, then they are usually ready to separate and store. Seeds need to be very dry before they are stored at temperatures below 95°F. The Organic Seed Alliance has very comprehensive information for those wishing to explore more in-depth information about seed saving.

☀ Appendix A
Hybrid Varities

BEANS
Bush

Derby
Goldencrop
Greencrop
Henderson Baby Bush
Improved Golden Wax
Jade
Roma II
Topcrop

Lima

Florida Speckled
Fordhook
Henderson

Pole

Landfrauen
Northeaster

BEETS
Pacemaker III
Red Ace
Ruby Queen

BROCCOLI
Emperor
Galaxy
Green Comet
Packman
Premium Crop

BRUSSELS SPROUTS
Jade Cross
Prince Marvel

CABBAGE
Golden Acre
Ruby Ball
Savoy Ace
Savoy King

CABBAGE, CHINESE
China Flash
China Pride
Jade Pagoda
Joi Choi
Mei Qing Choi
Monument

CANTALOUPE
Ambrosia
Caravelle
Explorer
Magnum 45
Mission
Perlita
Tam Uvalde

CARROT
Burpee's Toudo
Little Finger
Park's Nandor

CAULIFLOWER
Snow Crown

CHARD, SWISS
Bright Lights
Rhubarb Red

CORN
Frontier
Golden Queen

Guadalupe Gold
Honey and Pearls
How Sweet It Is
Kandy Korn
Merit Hybrid
Silver Queen
Sweet-G 90

CUCUMBER
Pickling Varieties
Calypso
Carolina

Slicing Varieties
Ashley
Burpless
Dasher II
Poinsett 76
Slicemaster
Sweet Slice
Sweet Success
Tasty Jade

EGGPLANT
Black Bell
Florida Market
Ichiban
Megal
Neon

GARLIC
Texas White

KALE
Blue Knight
Vates

LETTUCE
Buttercrunch
Tango
Winter Density

MUSTARD GREENS
Florida Broadleaf
Mizuna

OKRA
Baby Bubba
Cajun Delight
Lee
Louisiana Green Velvet

ONION
Beltsville (bunching type)
Crystal Wax (white)
Evergreen White (bunching type)

PEAS
Edible-Podded Peas
Snow Green
Sugar Bon
Sugar Mel
Sugar Pop

Southern (black-eyed)
California Blackeye #5
Cream 40
Zipper Cream

PEPPER
Hot Varieties
Garden Salsa
Hungarian Wax
Mexibell
Mucho Nacho
Numex Joe Parker
Slim Cayenne
Señorita Mild
Tabasco
Tam
Vera Cruz

Sweet Varieties
Big Bertha
Gypsy
Jingle Bells
Jupiter
Keystone Giant
Pimento L
Summersweet 860
Sweet Pickle
Yolo Wonder

POTATO, IRISH
Norland
Red La Soda
Russian Banana Fingerling

PUMPKIN

Aspen
Autumn Gold
Connecticut Field
Small Sugar
Spirit

RADISH

Early Scarlet Globe
Miyashige
Snow Belle

SPINACH

Coho
Fall Green
Hybrid 7
Melody
Samish
Space
Tyee

SQUASH

Summer

Butterstick
Dixie Yellow
Early Yellow Crookneck
Goldrush
Magda Zucchini
Multipik Yellow
Peter Pan
President Zucchini
Senator Zucchini
Sunburst Patty Pan

Sun Drops
White Bush Scallop

Winter

Acorn
Black Forest
Carnival
Cream of the Crop
Early Butternut
Sweet Dumpling
Sweet Mama
Table Ace
Table King Bush
Waltham Butternut

TOMATO

Cherry

Agriset
Juliet
Red Pear
Sugar Cherry
Sweet Baby Girl
Sweet Million
Sweet Olive
Yellow Pear

Roma

San Marzano
Viva Italia

Slicers

Better Boy
Burgess Stuffer

Carnival
Celebrity
Champion
Dona
Early Girl
First Lady
Jackpot
Merced
Quick Pick
Sunmaster
Superfantastic
Surefire
Yellow Stuffer
Whopper Improved

TURNIP

Royal Globe II
White Lady

WATERMELON

All Sweet
Orange Golden
Piñata
Royal Sweet
Star Brig

✳ Appendix B
Pest and Disease Control

Control of Fungal Diseases

The following organic fungal treatments may be specific to certain diseases. Make certain the product is labeled for the disease being treated. Fungal diseases will be more active in cool, rainy weather. Vigilance in keeping debris cleaned from bed areas, pruning dead or diseased plant material, and keeping air circulation open around healthy plant material will help keep fungal problems in check.

Actinovate: This highly effective fungicide is a relative newcomer to the organic arsenal. It is a high concentration of beneficial bacteria effective in controlling a wide variety of common plant fungal diseases, including powdery and downy mildew and diseases caused by fungi in the genera *Pythium, Rhizoctonia, Fusarium, Phytophthora, Verticillium, Botrytis,* and *Alternaria*. It also contains *Streptomyces lydicus*, a beneficial soil microorganism. This is a very exciting product for organic gardeners.

Baking soda: This is a home remedy for fungal diseases on plants. Simply mix 4 teaspoons of baking soda in a gallon of water. The solution will stick to leaves better if a couple of drops of dish soap or vegetable oil are added. Shake well and spray on leaves.

Bordeaux mixture: This traditional fungicide is a mixture of hydrated lime and copper sulfate suspended in water and sprayed on the leaves of plants. Bordeaux mixture should not be applied when temperatures are above 85°F.

Neem oil: This horticultural oil is derived from the neem tree and has myriad uses. The oil is an insecticide for small sucking insects such as aphids and whiteflies. It is also labeled for the control of black spot, anthracnose, powdery mildew, and rust. This oil can burn foliage if used at temperatures above 85°F. It is usually best to spray in the early morning or evening when temperatures are cooler.

Serenade: Strains of *Bacillus subtilis* are the active ingredient in this fungicide that is used to treat black spot, powdery mildew, rust, gray mold, late blight, and scab.

Insect Problems

Pyrethrum is an effective organic pesticide that is still on the market. It is a derivative of a type of chrysanthemum. Even though it is safe for humans, it has been determined to be toxic to fish and beneficial insects. It should be avoided unless absolutely necessary for this reason. Rotenone is another insecticide that is a botanical derivative. It is even more toxic to fish and has been linked to Parkinson's disease in rats. Rotenone is used as a piscicide to kill undesirable fish species in lakes before they are restocked. Rotenone's potential dangers should prevent its use as an insecticide in the edible garden.

Bacillus thuringiensis (**Bt**): This beneficial bacteria that can be applied as a powder or a spray when mixed with water primarily to kill caterpillars and larvae of harmful chewing insects. *B. t.* var. *israelensis* is used to kill mosquito larvae in ponds. *B. t.* var. *san diego* is used to control the Colorado potato beetle.

Beauveria bassiana: This biological fungus has shown varying degrees of efficacy against the European corn borer, grasshoppers, tarnished plant bugs, and whiteflies, among others. The fungal spores have to come in direct contact with the insects, and some formulations break down in high temperatures. Repeated applications offer better control.

BENEFICIAL INSECTS

There are many types of beneficial insects. Natural populations can be encouraged when chemical pesticides and herbicides are avoided. The following are the more common beneficial insects, available at local nurseries or by mail order.

Beneficial nematodes: These microscopic predators are wormlike in appearance. There are many types of nematodes. Some are harmful, and some are beneficial. Types of beneficial nematodes attack the root-knot nematode, which is a common pest for carrots and other root crops. These beneficial nematodes are sold in packets. The contents are mixed with water and sprayed on the soil to attack larvae of fleas, mole crickets, Japanese beetles, and weevils, as well as grubs and fire ants. Applications are usually made in the spring after the weather has warmed and should be applied to wet soil.

Green lacewing: This adult insect's delicate lacy wings and small appearance give no clue to the terror it invokes in its larval stage when feeding on aphids, red spider mites, thrips, and mealybugs.

Ladybug (**ladybird**) **beetles:** The larval and adult stages of this insect are voracious predators of aphids, scale, thrips, and mealybugs. It is best to release them in the evening onto infected plants.

Praying mantids: These predator insects appear to be in a meditative trance as they lie in wait for unsuspecting caterpillars, grasshoppers, and beetles.

Spiders: Despite their reputation, spiders are very helpful in controlling insects that are harmful. The majority of spiders encountered in the garden in Texas pose no threat to humans. They are not generally available through garden centers or mail order.

Trichogramma wasps: These tiny wasps are about the size of a gnat and are harmless to humans. They lay their eggs in the eggs or larvae of other species. The larvae hatch and parasitize the host. They are used to control pecan case bearers, cabbage loopers, tomato hornworms, and corn earworms.

Wasps: Wasps are beneficial. Because of their painful sting and sometimes allergic reactions caused in humans, they have been seen as a harmful pest; however, if there is no danger of allergic reaction, they can be very helpful in controlling caterpillars, hornworms, codling moths, and aphids.

CITRUS OIL

The active ingredient in citrus oil is limonene, which is extracted from the rinds of citrus fruit. Citrus oil acts on the nervous system of pests that afflict animals, such as fleas, lice, ticks, and mites. The oil is nontoxic to mammals. Use citrus oil with care. It can burn and desiccate plants in high temperatures.

DIATOMACEOUS EARTH (DE)

This white, chalky powder is a powdered form of the fossils of diatoms, a type of hard-shelled algae. The powder is an abrasive to the exoskeletons of a wide variety of insects, including coddling moths, twig borers, thrips, mites, slugs, snails, aphids, and ants. The product must come in contact with the insects to be effective. DE can be applied as a dust or as a wettable powder. Wear a mask when applying the dust, as it is an irritant to mucous membranes in the nose. It is also used to filter pool water. This type of DE is not good for gardens. Be sure to buy the product labeled "food grade" or "insecticidal."

HORTICULTURAL OIL

This mineral oil–based insect control blocks the holes through which insects breathe and causes them to suffocate. The oils are used to control aphids, mites, leafhoppers, whiteflies, and scale. Some of these products can cause phytotoxicity or sunscald if they are sprayed in temperatures above 85°F, especially dormant oils. Dormant oils should be applied only before bud break. Horticultural oils are good surfactants for baking soda fungal controls, helping the product stick to the leaves. They should not be used with sulfur applications or within 30 days of sulfur applications on plant leaves.

INSECTICIDAL SOAP

Soap-based insecticides are not residual, so they must come in contact with the insect to be effective and may require repeated applications. The soap is used to control soft-bodied arthropods, including aphids, young scale, whiteflies, psyllids, mealybugs, and spider mites. It is believed that insecticidal soaps may affect the insect's protective waxy coatings. Some plants, such as plums and tomatoes, are susceptible to damage from soap-based products. Be careful to read the label to make sure the insecticide is labeled for use on infested plants.

NEEM OIL

Neem oil is a systemic insecticide that the plants take up deep inside their tissues. Neem is a hormonal agent that works on chewing and sucking insects that have mouthparts that can reach the inner parts of the leaves, such as aphids, spider mites, scale, whiteflies, beetles, and leaf rollers. It is thought that neem oil acts as a repellent and may also interfere with the insects' natural hormonal activity that tells them to eat and mate. Obviously, if the insects do not eat or mate, it will not take long for the population to die out. Neem oil is also used as an organic fungicide for a number of fungal problems, including powdery mildew, black spot, downy mildew, anthracnose, rust, leaf spot, botrytis, and alternaria.

NOSEMA LOCUSTAE

This insecticide is used to control grasshoppers. It is a single-celled microsporidium protozoan that infects the grasshoppers and spreads to the rest of the population as they cannibalize the infected grasshoppers. *N. locustae* should be applied when grasshoppers are small for the most effective control. This has been shown to be a very effective control against grasshoppers, but it will take a few days to over a week to take effect. A second application in 4 to 6 weeks will be required if grasshoppers are still present. Some carryover of insecticide will last into the next year.

PYRETHRUM (PYGANIC PRO)

This insecticide is derived from a type of chrysanthemum. It is one of the few organic insecticides labeled for the control of beetles, as well as many other garden pests. However, this insecticide is broad spectrum and will kill beneficial insects as well as harmful ones. It is also harmful to fish and should not be used near water features. This product should be used sparingly and as a last resort to control severe infestations. It is labeled for about one hundred insect pests, including hard-to-kill beetles.

SPINOSAD

This bacterial by-product of fermentation was approved for organic use in 2003. The bacteria (*Saccharopolyspora spinosa*) are ingested by leaf-eating insects. Spinosad is used to control caterpillars, beetles, and thrips. It is one of the few products (organic or chemical) that effectively controls beetles. It persists longer than Bt for caterpillar control (4 weeks rather than 2 weeks for Bt). Spinosad is also labeled for fire ant control for commercial applicators. Entrust is the wettable powder labeled for use on edible plants. It is safe for humans; however, bees and butterfly larvae are susceptible to harm. It should not be sprayed when bees are foraging or near butterfly larvae plants, such as fennel or parsley.

SULFUR

Sulfur is an excellent miticide that can be dusted on ankles to prevent chiggers. It can also be used to control red spider and other mites. Be careful not to breathe the dust, as it is an irritant. Sulfur should not be applied when temperatures are above 85°F. (Unfortunately, these are the temperatures when spider mites are most active.)

SURROUND

This fine, kaolin clay is suspended in water and sprayed on fruit crops commercially to protect against sunburn and heat stress. It also protects against plum curculio, leafhoppers, aphids, Japanese beetles, crickets, thrips, grasshoppers, white-flies, cutworms, weevils, and loopers. There will be a dusty, white film on the plants after application that can be washed off easily before eating. It is best to apply Surround when the foliage is wet for better adhesion. Multiple applications may be required if the product is washed off by rain. Be sure to coat both sides of the leaves.

Disease and Insect Symptoms and Treatments

Although many insect and disease problems occur in all plant families, some are specific to a particular family. The following tables group these symptoms by family and provide organic treatment methods.

Insect and disease problems of pecan trees		
Insect/disease	*Symptoms*	*Organic treatment*
Aphid	Small, soft-bodied, oval-shaped insect that causes sticky sap to fall on objects below trees.	Release beneficial insects such as ladybeetles and green lacewings. Spray with insecticidal soap, neem, or orange oil. Select resistant varieties.
Pecan nut casebearer	White, pink, or red eggs at the nut tip. Small webbing and black excrement (frass) at the tip of the nut. Misshapen and ill-formed nuts.	Release trichogramma wasps. Apply spinosad or Bt (*Bacillus thuringiensis*).
Webworm	Large mass of webbing in trees.	Break open webs to allow natural predators to attack. Spray Bt into webs.
Zinc deficiency	Crinkled, bronzed leaves and bare branches or clusters of leaves at the tips of otherwise bare branches.	Apply zinc as a foliar feed if it will not be sprayed on other desirable plants. Apply zinc in dry form at the base of the tree.

Insect and disease problems of stone fruit trees

Insect/disease	Symptoms	Organic treatment
Bacterial spot	Small, watery lesions on underside of leaves that become larger and can cause leaves to yellow. Cracked fruit that may form brown and black spots.	Remove infected leaves and fruit and keep debris cleaned from under trees. Prune trees to allow good air circulation. Apply Actinovate or Serenade if infection occurs.
Brown rot	Brown spots on blossoms and white circular spots on fruit.	Remove debris from around tree and prune to allow good air circulation. Apply Actinovate if infection occurs.
Catfacing bug	Gnarled fruit.	Remove debris from around tree. Apply kaolin clay product to help prevent infestation and pyrethrum products for heavy infestations.
Leaf curl	Thick, puckered leaves that turn reddish and fall about a month after bloom.	Usually not a severe problem. Remove diseased leaves and fruit. Apply Bordeaux mixture during late dormancy as a preventive.
Oriental fruit moth	Stunted terminal growth in leaf formation.	Prune damaged fruit and leaf clusters and put in trash. Release trichogramma wasps.
Peach tree borer	Gummy substance at the base of the tree.	Clean gummy substance from tree and use a rigid wire in the hole to kill borers. Spray parasitic nematodes (*Steinernema carpocapsae*) at base of tree.
Plum curculio	Small, gray, long-snouted beetle that feeds on leaves, flowers, and fruit.	Remove debris from around tree. Apply kaolin clay product to help prevent infestation. Treat with spinosad or Bioganic.

Insect and disease problems of apple trees

Insect/disease	Symptoms	Organic treatment
Apple scab	Dark brown, velvety brown to olive spots on fruit and leaves.	Keep area around tree free of debris. Apply Actinovate or Serenade if problem develops.
Black rot	Sunken reddish lesions on tree bark and fruit. Lesions on leaves, flaking bark, and end rot on fruit.	Keep area around tree free from debris.
Cedar apple rust	Yellow spots on the tops of leaves. White, tubular structures growing from leaves or fruit.	Inspect local cedar trees for galls and bright orange spore horns. Remove and dispose of these growths. Remove and dispose of infected fruit and leaves.
Codling moth	Grayish-brown moth with fringe-tipped wings. Holes in fruit. Larvae in fruit.	Keep area around tree free from debris. Maintain organic environment to encourage natural predators.
Cotton root rot	Brown fungal hyphal strands on roots. Overall decline and death of tree. More common in alkaline soils.	No control. Check for pathogen with soil test before planting. Remove trees and do not replant in area.
Fire blight	Blossoms and terminal leaf growth turn brown, wither, and die. Common "shepherd's crook" withering of terminal growth.	Remove limbs at least 8 inches below damage with sterilized pruners.
Phytophthora collar rot	Slow bud formation. Leaf discoloration and twig dieback. Water molds that appear as brownish-red rot near the graft union.	Choose resistant rootstock. Keep dead and diseased wood and fruit cleaned from around the tree. Apply Serenade.
Woolly apple aphid	Long, white, waxy filament on leaves and bluish-white, rod-shaped secretions on underground roots.	Choose resistant rootstock. A tiny wasp, *Aphelinus mali*, is a natural predator.

Insects and diseases of citrus trees

Insect/disease	Symptoms	Organic treatment
Aphid	Small, pear-shaped, soft bodied insect that attacks leaves, especially new growth, causing stunting and deformation.	Release beneficial insects. Use strong spray of water from a hose, insecticidal soap, or neem or orange oils.
Asian citrus leaf miner	Serpentine trails in stunted and misshapen leaves.	Spray with citrus oils at first signs of disease. Release parasitic wasps, spiders, or green lacewings.
Citrus canker	Yellow and gray blotchy lesions on fruit and leaves. Fruit and leaf drop. Not currently in Texas, due to a quarantine of citrus from other states.	Order trees only from Texas nurseries. Report disease immediately to local county extension agent.
Citrus flatid leafhopper	Grayish-white insect about 3/8-inch long. Nymphs secrete a white, waxy material. Honeydew secretions on leaves.	Controls usually not necessary.
Citrus greening	Mottling on older leaves; yellowing of leaves in sectors. Infected fruit small and misshapen and has a bitter taste. Seeds are black.	Order citrus trees only from reputable nurseries. Recently found in Texas. If citrus greening suspected, report immediately to county extension agent. Control psyllid vector by applying pyrethrum, dormant oils, neem oil, or kaolin clay and destroy infected trees.
Greasy spot	Yellow spots on leaves, turning dark and greasy. Rind blemishes.	Remove affected leaves and fruit. Apply Serenade.
Melanose	Circular dark depressions with yellow margins on twigs and leaves. Spots become raised and dark.	Remove the most seriously affected areas.
Orange dog caterpillar	Chewed leaves. Mottled gray and brown caterpillars become swallowtail butterflies.	Controls usually not necessary. Can be handpicked or heavy infestations controlled with Bt (*Bacillus thuringiensis*).
Phytophthora collar rot (foot rot)	Brownish-red water mold near the graft union, on the scion side. Slow bud formation, leaf discoloration, and twig dieback.	Select resistant rootstock varieties. Apply Actinovate as soil drench.
Puss caterpillar	Chewed leaves. Fuzzy tan or gray caterpillars with venomous sting.	Controls usually not necessary. Heavy infestations can be controlled with Bt.
Scab	Translucent dots that swell to corky, wartlike bumps, especially on new growth. Twisted, puckered leaves.	Remove the most seriously affected areas and apply Actinovate or Serenade.
Sooty mold	Black, sootlike mold on leaves and fruit.	Usually associated with aphids or other honeydew-secreting insects. Control aphids and remove infected leaves and fruit.
Twig dieback	Small, round, whitish blisters that turn tan or brown with a yellow halo. Oily area around lesions.	Contact county extension agent. If it is twig dieback, the tree must be destroyed.

Insect and disease problems of loquat trees

Insect/disease	Symptoms	Organic treatment
Aphid	Small, soft-bodied, oval-shaped insect that causes sticky sap to fall on objects below trees.	Release beneficial insects such as ladybeetles and green lacewings. Spray with insecticidal soap, neem, or orange oil.
Black scale	Black or brown, hard-shelled insect that attaches to leaves and produces a honeydew that can cause sooty mold.	Scratch insects off with fingernail and drop into alcohol or cooking oil. Apply horticultural oil in winter as a preventive.
Fire blight	Blossoms and terminal leaf growth turn brown, wither, and die. Common "shepherd's crook" withering of terminal growth. Dark brown, shriveled fruit.	Remove limbs at least 8 inches below damage with sterilized pruners.
Phytophthora collar rot	Slow bud formation. Leaf discoloration and twig dieback. Water molds that appear as brownish-red rot near the graft union.	Treat with Serenade.

Insect and disease problems of pear trees

Insect/disease	Symptoms	Organic treatment
Cotton root rot	Brown fungal hyphal strands on roots. Overall decline and death of tree. More common in alkaline soils.	No control. Check for pathogen with soil test before planting. Remove trees and do not replant in area.
Fire blight	Blossoms and terminal leaf growth turn brown, wither, and die. Common "shepherd's crook" withering of terminal growth. New growth shriveled.	Remove limbs at least 8 inches below damage with sterilized pruners.
Pear psylla	Small sucking insect similar to an aphid that spreads disease. The honeydew they produce increases the likelihood of sooty molds.	Apply kaolin clay product to help prevent infestation.

Insect and disease problems of fig trees

Insect/disease	Symptoms	Organic treatment
Root-knot nematodes	Declining health in plants. Galls or knots on roots.	Inspect roots before planting. Add active humus or compost tea around the roots. Apply ClandoSan.
Souring	Beetles enter the fruit through the "eye" and ruin the fruit.	Select resistant varieties: 'Alma,' 'Celeste,' and 'Texas Everbearing.'

Insect and disease problems of asparagus

Insect/disease	Symptoms	Organic treatment
Aphid	Small, soft-bodied, oval-shaped insect that causes sticky sap to fall on objects below trees.	Release beneficial insects such as ladybeetles and green lacewings. Spray insecticidal soap, neem oil, or orange oil. Select resistant varieties.
Asparagus beetle	Brightly colored, red-headed beetle with spotted red or black body that feeds on shoots, causing distortion and scarring.	Keep debris cleaned from around base of plants where beetles overwinter.
Asparagus rust	Yellow-orange blisters, starting on the leaves (or needles) and spreading to the rest of the plant.	Select resistant varieties. Apply Serenade or neem oil.
Fusarium wilt	Stunted, brown spears. The internal areas of the crowns turn yellowish-red, and the roots rot.	Carry out good bed preparation and sanitation. Apply Actinovate as a soil drench.
Needle blight	Fronds develop buff to gray elliptical spots with a purple halo and lose their needles. More common during rainy season.	Thin plants for increased air circulation and removal of debris. Apply Serenade.
Purple spot	Sunken purple spots on the spears.	Clean fallen leaf debris out of the beds before new growth emerges.

Insect and disease problems of bramble fruits

Insect/disease	Symptoms	Organic treatment
Anthracnose	Small, purplish spots on new shoots. New canes may be stunted. Small, scabby fruit.	Remove diseased fruit. Make sure fruit is not in contact with soil and has plenty of air circulation. Water with drip systems. Apply Serenade.
Cane borer	Red-headed insect with long, slender black body that bores into canes. Canes weaken and wilt and eventually die.	Check for small holes in canes and cut off infected canes. Do not compost infected canes.
Crown gall	Swelling at the base of the plant.	Remove infected plants and buy plants from reputable nurseries.
Orange rust	Masses of orange spots on leaves.	Remove and destroy infected plants.
Spider mite	Lighter leaf color. Tiny webbing on the underside of leaves.	Remove and destroy all infected leaves. Spray all sides of leaves with strong spray of water. Apply spinosad or garlic pepper tea.
Stink bug (leaf-footed bug)	Shield-shaped insect that sucks juices from fruit. Discolored fruit with hard blisters.	Inspect underside of leaves for eggs and destroy if found. Clean debris from around canes. Apply Surround WP as preventive and PyGanic Pro as last resort.
Strawberry root weevil	Small, shiny, black adults chew on leaves; the larvae chew on roots.	Keep debris cleaned from around plant. Apply Surround. Release parasitic nematodes (*Heterorhabditis bacteriophora*).

Insect and disease problems of grapes

Insect/disease	Symptoms	Organic treatment
Anthracnose	Lesions, especially on shoots and fruit, beginning as red, circular spots and enlarging to sunken, circular, or angular, gray lesions. Edges may be black and raised. Reddish, circular lesions on fruit have gray centers.	Remove and destroy diseased parts of plants. Make sure debris is cleaned from bed and prune vines to allow good air circulation. Water with drip systems. Apply Serenade or Trilogy.
Black rot	Brown, circular lesions on leaves. Purple to black, sunken lesions on stems and tendrils. Fruit turns light brown and then turns black and withers.	Sanitation is key in control. Remove infected fruit from vines and destroy.
Bunch rot	Dull green spots near leaf veins. Blighting of blossoms and shoots. Grape clusters covered with beige, dusty mold and rot.	Keep plant debris cleaned from beds. Remove infected fruit from vines and destroy. Apply Serenade.
Cotton root rot	General decline of the plant's health, sometimes over a period of years. Foliage discolored and plants wilt as disease progresses.	No treatment available. Test soil for pathogen before planting.
Downy mildew	Rust and yellow mottling on upper side of leaves. Fuzzy, gray spots on underside. Generally a late-season disease.	Remove infected leaves and fruit and destroy. Apply Serenade, Trilogy, or Actinovate.
Pierce's disease	Dries and causes browning of leaves and necrosis of vines.	Spread by the glassy-winged sharpshooter insect. Plant resistant varieties. All varieties in this book are resistant.
Powdery mildew	Light, blotchy, gray spots on leaves and stunted fruit growth. Most common in cooler, rainy months.	Remove infected leaves and fruit and destroy. Apply Serenade, Trilogy, or Actinovate.

Insect and disease problems of strawberries

Insect/disease	Symptoms	Organic treatment
Gray mold botrytis	Shrunken, gray fruit.	Remove infected fruit, and make sure there is mulch between the fruit and the soil. Apply Actinovate, Serenade, or neem oil.
Nematodes	Wilted, stressed plants that appear to need water.	Add well-decomposed organic matter to soil before planting. Plant cover crops in the off season. Apply compost tea or ClandoSan.
Verticillium wilt	Outer leaves droop, wilt, turn dry, and become reddish-yellow or dark brown at the margins and between veins.	Do not plant in the same bed where tomatoes, eggplant, or potatoes were grown in the previous year. Apply Actinovate as a soil drench.
Western tarnished plant bug	Small beetle with green or yellow "V" at the base of the head that causes puckered, inedible fruit.	Remove plant debris from bed. Release parasitic wasps (*Anaphes iole*).

Insect and disease problems of beet family

Insect/disease	Symptoms	Organic treatment
Aphid	Small, soft-bodied insect, usually on new growth, that can cause leaf discoloration and stunt new growth. Can also produce large quantities of a sticky exudate known as honeydew, which often turns black with the growth of a sooty mold fungus.	Release beneficial insects such as ladybeetles and green lacewings. Apply insecticidal soap, neem, or orange oil.
Beet armyworm	Small, green worms emerge from cottony masses on leaves and feed on terminal growth.	Remove leaves with cottony masses and destroy. Release parasitic wasps. Treat with Bt (*Bacillus thuringiensis*).
Boron deficiency	Necrosis of new growth, leaf discoloration, and wilting. Can resemble symptoms of other nutritional deficiencies.	Add organic matter. Test for boron in soil and treat as needed.
Cabbage looper	Gray-brown moth with "V" on forewings, active at night. Green "inchworms" (the larvae) eat ragged holes in leaves.	Cover plants with floating row covers at night. Apply Btk (*B. t.* var. *kurstaki*) or spinosad.
Cercospora leaf spot	Tan to yellowish spots on leaves turn necrotic, with a wet, dark, and usually sunken and/or translucent appearance as if water soaked.	Remove infected leaves and destroy. Use drip irrigation and ensure good air circulation. Apply Serenade.
Cucumber beetle	Yellow beetle with black stripes or dots that causes interveinal webbing of and holes in leaves in amaranth plants.	Encourage natural predators. Remove infected plants and debris from garden bed.
Downy mildew	Rust and yellow mottling on upper side of leaves. Fuzzy, gray spots on underside. Generally a late-season disease.	Remove infected leaves and destroy. Apply Serenade, neem products, or Actinovate.
Flea beetle	Small beetles jump when leaves are brushed and bore tiny holes in leaves.	Keep debris cleaned from area. Apply garlic tea. Apply beneficial nematodes to attack larvae. Apply neem (pyrethrum products as a last resort).
Leaf miner	White, winding tunnels in leaves.	Remove infested leaves. Apply spinosad (will also harm butterfly larvae and bees).
Nematodes, root-knot	Wilted, stressed plants are lighter and appear to need water.	Add well-decomposed organic matter to soil before planting. Plant cover crops in the off season. Rotate crops. Apply compost tea or ClandoSan.
Powdery mildew	Light, blotchy gray spots on leaves. Most common in cooler, rainy months.	Remove infected leaves and destroy. Apply Serenade, neem products, or Actinovate.
Pythium	Seedlings rot off at soil line.	Water in seedlings with fish emulsion and seaweed at planting. Make sure beds are well drained. Apply Actinovate.
Viruses (cucumber mosaic virus, beet curly top, beet western yellows)	Discoloration of leaves that starts in one part of plant and spreads. Wilting and foliage loss.	Spread by insects such as aphids and leafhoppers. Control insects. Remove and destroy infected plants.
White rust	Small, yellow spots on upper side of leaves become glassy, white pustules.	Remove infected leaves and destroy. Use drip irrigation and ensure good air circulation.

Insect and disease problems of carrots

Insect/disease	Symptoms	Organic treatment
Alternaria leaf blight	Dark, irregular lesions on leaves.	Buy seed from reputable sources. Apply Actinovate or Serenade.
Aster yellows	Stunted, yellow leaves.	Control leafhoppers. Remove infected plants. Keep weeds controlled in and around garden.
Carrot weevil	Small, dark to copper-colored insect whose white grub larvae burrow into carrot roots.	Clean leaf debris and weeds from bed and surrounding area.
Nematodes, root-knot	Wilted, stressed plants that appear to need water.	Add well-decomposed organic matter to soil before planting. Plant cover crops in the off season. Rotate crops. Apply compost tea or ClandoSan.
Powdery mildew	White or gray "dust" covers plant.	Remove infected leaves. Apply Actinovate, Serenade, baking soda, or neem product.
Southern blight	Carrot roots rot. White fungal mats.	Turn soil well before planting, make sure soil drains well, and rotate location of crop families.
Wireworm	Shiny reddish-brown to yellow wormlike insect that feeds underground and burrows into carrot roots.	Turn soil well before planting to expose larvae and rotate crop locations. Plant cover crops,. Release beneficial nematodes. Incorporate wood ashes. Destroy infected carrots.

Insect and disease problems of corn

Insect/disease	Symptoms	Organic treatment
Armyworm	Fuzzy white egg masses on the underside of leaves. Thinning or full holes in leaves. Brownish caterpillars 1½ to 2 inches long with longitudinal stripes feed at night.	Apply Bt (*Bacillus thuringiensis*).
Corn earworm	Caterpillars 1½ inches long that are brown to green or may have a light pink color with longitudinal stripes burrow into ear and eat kernels.	Select varieties with tighter ears. Inspect silks for eggs. Apply Btk (*B. t.* var. *kurstaki*) with vegetable oil to silks before larvae enter ears but after silks have started to brown.
Corn leaf aphid	Bluish-green, small, soft-bodied insect usually on new growth and maturing silks and tassels that can cause leaf discoloration and puckered growth. Also produces large quantities of a sticky exudate known as honeydew, which often turns black with the growth of a sooty mold fungus.	Release beneficial insects such as ladybeetles or green lacewings. Apply a strong spray of water from a hose. Apply garlic pepper tea. Keep weedy grasses away from garden and surrounding area.
Corn rootworm	Larvae attack roots and cause them to topple. Adults are yellow beetles with black spots and feed on leaves and silks.	Natural predators include bats, wolf spiders, and braconid wasps. Use sticky traps. Remove infected plants and destroy. Apply pyrethrum as a last resort for control of adults.
Corn smut	Grayish, puffy fungal growth in ears.	Carefully collect infected ears and put in the trash. Be careful not to break open the fungal masses.
Cutworm	Young seedlings cut at ground level.	Keep garden free of debris. Release beneficial nematodes or trichogramma wasps. Apply Btk, diatomaceous earth, or *B. subtilis* (Serenade).
European corn borer	Whitish-yellow egg masses on the underside of leaves. Light pink or gray larvae with brown spots on each segment burrow into ears and stalks, causing the plant to fall over.	Remove all corn stalks and ears from garden after harvest. Release trichogramma and braconid wasps, assassin bugs, mantids, and spiders. Apply the fungus *Beauveria bassiana*.
Mosaic virus	Mottled yellow markings on leaves.	Spread by corn leaf aphid. Control aphids and keep weeds away from surrounding area.
Rust	Reddish lesions and pustules on the upper and lower leaf.	Use drip irrigation. Ensure good air circulation. Remove infected leaves and keep debris clean from beds. Apply *B. subtilis*.

Insects and diseases problems of crucifers

Insect/disease	Symptoms	Organic treatment
Alternaria leaf spot	Yellow or dark brown to black lesions on leaves.	Buy seeds from reputable sources. Apply Actinovate or Serenade.
Anthracnose	Dry, gray to straw-colored, circular lesions on leaves; lesions on stems gray to brown with black borders.	Provide good air circulation, and remove infected leaves. Apply Serenade.
Aphid	Small, soft-bodied insect, usually on new growth, that can cause leaf discoloration and stunt new growth. Can also produce large quantities of a sticky exudate known as honeydew, which often turns black with the growth of a sooty mold fungus.	Release beneficial insects such as ladybeetles and green lacewings. Apply insecticidal soap, neem, or orange oil.
Black leg	Black lesions on newly formed leaves. Roots turn black and rot.	Buy certified disease-free seeds. Plant in well-drained soil. Remove infected plants. Do not replant in area for a few years.
Black rot	Yellow lesions extend toward base of leaf. Blackened veins on roots.	Remove infected plants and destroy. Do not replant in area for a few years.
Boron deficiency	Hollow stems in broccoli. Generally poor development.	Add organic matter. Test for boron in soil and treat as needed.
Cabbage looper	Gray-brown moth with "V" on forewings; active at night. Green "inchworms" (the larvae) eat ragged holes into leaves.	Cover plants with floating row covers at night. Apply Btk (*Bacillus thuringiensis* var. *kurstaki*) or spinosad.
Cabbage root maggot	Small flies lay eggs at the base of plants. Larvae eat the lateral roots then the taproot. Plant turns purplish or yellow, withers, and dies.	Plant clover crops. Apply a ring of diatomaceous earth and/or wood ashes around the base of the plant. Cover plants with floating row covers.
Club root	Swollen roots and yellowing of plants.	Carefully dig up and destroy infected plants. Do not replant in area for a few years. Plant cereal rye as cover crop.
Diamondback moth	The ½-inch green caterpillars (larvae of gray moth with "V" on back) chew leaves.	Cover plants with floating row covers at night. Apply Btk or spinosad.
Downy mildew	Rust and yellow mottling on upper side of leaves. Fuzzy, gray spots on underside. Generally a late-season disease.	Remove infected leaves and destroy. Apply Serenade, Trilogy, or Actinovate.
Flea beetle	Small beetles jump when leaves are brushed and bore tiny holes in leaves.	Keep debris cleaned from area. Apply garlic tea, neem, or pyrethrum products.
Fusarium wilt	Yellowing of outer leaves, commonly on one side.	Plant resistant varieties. Apply Actinovate.

Harlequin bug	Orange and black shield-shaped beetle that sucks juices from plants.	Keep beds free of debris. Apply pyrethrum as last resort
Imported cabbage worm	Green caterpillars about an inch long with a yellow stripe down their backs feed on leaves. (The adult is a white butterfly about 2 inches across.)	Cover plants with floating row covers at night. Apply Btk or spinosad.
Head rot	Water-soaked lesions on the leaves become soft spots and are covered with a fuzzy, white growth.	Keep beds and surrounding area free from weeds. Do not plant in beds that have been previously infected. Apply Serenade.
Nematodes, root-knot	Wilted, stressed plants are lighter and appear to need water.	Add well-decomposed organic matter to soil before planting. Plant cover crops in the off season. Rotate crops. Apply compost tea or ClandoSan.
Powdery mildew	Light, blotchy gray spots on leaves. Most common in cooler, rainy months.	Remove infected leaves and destroy. Apply Serenade, Trilogy, or Actinovate.
Pythium	Seedlings rot off at soil line.	Water in seedlings with fish emulsion and seaweed at planting. Make sure beds are well-drained. Apply Actinovate.
Rhizoctonia	Roots, stems, and leaves rot.	Remove infected plants and treat with Actinovate.
Thrips	White blotches and streaking on leaves.	Release green lacewings and ladybeetles. Apply spinosad, neem, or insecticidal soap.

Insect and disease problems of cucurbits

Insect/disease	Symptoms	Organic treatment
Aphid	Small, soft-bodied insect, usually on new growth, that can cause leaf discoloration and stunt new growth. Can also produce large quantities of a sticky exudate known as honeydew, which often turns black with the growth of a sooty mold fungus.	Release beneficial insects such as ladybeetles and green lacewings. Apply insecticidal soap, neem, or orange oil.
Bacterial wilt and cucumber mosaic virus	Dull green patches on leaves. Wilting spreads laterally. Blotches or mottled light yellow patterns on the leaves may be followed by leaf and fruit stunting.	Keep beds free of debris. Remove diseased leaves. Control striped cucumber beetles and spotted cucumber beetles. Use floating row covers. Release braconid wasps. Apply beneficial nematodes (*Steinernema* spp. and *Heterorhabditis* spp.). Build bat houses to attract bats. Plant amaranth as catch crop. Treat with pyrethrum products as last resort.
Downy mildew	Rust and yellow mottling on upper side of leaves. Fuzzy, gray spots on underside. Generally a late-season disease.	Remove infected leaves and destroy. Apply Serenade, Trilogy, or Actinovate.
Leaf miner	White, winding tunnels in leaves.	Remove infested leaves. Apply spinosad (will also harm butterfly larvae and bees).
Nematodes, root-knot	Wilted, stressed plants are lighter and appear to need water.	Add well-decomposed organic matter to soil before planting. Plant cover crops in the off season. Rotate crops. Apply compost tea or ClandoSan.
Pickleworm	Adult moths have dark brown wing margins with "tufts" at the base of the abdomen. Larvae are yellow-green with black spots and feed on the blossoms, tunneling into the fruit. Stems wilt and fruit dies.	Plant early-maturing varieties. Use floating row covers. Treat with Bt (*Bacillus thuringiensis*), which is effective before larvae burrow. Release beneficial nematodes (*Steinernema carpocapsae*). Remove the chrysalis from leaves.
Powdery mildew	Light, blotchy gray spots on leaves. Most common in cooler, rainy months.	Remove infected leaves and destroy. Apply Serenade, Trilogy, or Actinovate.
Spider mite	Lighter leaf color. Tiny webbing on the underside of leaves.	Remove and destroy all infected leaves. Spray all sides of leaves with strong spray of water. Apply spinosad or garlic pepper tea.
Spotted cucumber beetle	Yellow beetle with black spots that feeds on leaves, stems, and fruit. Larvae feed on roots and stems.	Spreads bacterial wilt and cucumber mosaic virus. Handpick beetles from plants. Keep beds free of debris. Treat with pyrethrum products.

Squash bug	Dark brown to gray insect with "V" at base of the head. Adults and nymphs cause damage by sucking plant juices. Reddish eggs on the underside of leaves. Plants wilt and turn black.	Inspect underside of leaves daily. Keep debris cleaned from beds.
Squash vine borer	Frass at the entry of small holes in stems. Greenish-brown moth with orange-red markings. Larvae are white grubs with brown heads. Brownish, flat eggs on the underside of leaves. Larvae cause damage by boring into stems, blocking the flow of water within the plant.	Inspect underside of leaves daily. Slit stems and remove larvae. Wrap pantyhose around lower stems. Keep debris cleaned from beds. Use floating row covers when plants not pollinating. Inspect soil often for larvae. Inject Bt into infected stems.
Striped cucumber beetle	Yellow beetle with black stripes that feeds on leaves, stems, and fruit. Larvae feed on roots and stems.	Spreads bacterial wilt and cucumber mosaic virus. Handpick beetles from plants. Keep beds free of debris. Treat with pyrethrum products.
Whitefly	Clouds of small white insects fly into the air when leaves are brushed. Stunted growth. Sticky honeydew on leaves.	Apply neem or orange oils if temperatures are below 85°F. Seaweed foliar spray has some efficacy.

Insect and disease problems of legumes

Insect/disease	Symptoms	Organic treatment
Aphid	Small, soft-bodied insect, usually on new growth, that can cause leaf discoloration and stunt new growth. Can also produce large quantities of a sticky exudate known as honeydew, which often turns black with the growth of a sooty mold fungus.	Release beneficial insects such as ladybeetles and green lacewings. Apply insecticidal soap, neem, or orange oil.
Cercospora leaf spot	Tan to yellowish spots on leaves turn necrotic and give a wet, dark, and usually sunken and/or translucent appearance to the affected area that looks water soaked.	Remove infected leaves and destroy. Use drip irrigation. Ensure good air circulation. Apply Serenade.
Downy mildew	Rust and yellow mottling on upper side of leaves. Fuzzy, gray spots on underside. Generally a late-season disease.	Remove infected leaves and destroy. Apply Serenade, Trilogy, or Actinovate.
Fusarium wilt	Yellowing of older leaves.	Plant resistant varieties. Apply Actinovate.
Mosaic virus	Light and dark green mosaic patterns with puckering and distortion on leaves	Plant resistant varieties. Keep beds free of debris and destroy infected plants. Control aphids and other insect vectors. Rotate crops.
Nematodes, root-knot	Wilted, stressed plants are lighter and appear to need water.	Add well-decomposed organic matter to soil before planting. Plant cover crops in the off season. Rotate crops. Apply compost tea or ClandoSan.
Powdery mildew	Light, blotchy gray spots on leaves. Most common in cooler, rainy months.	Remove infected leaves and destroy. Apply Serenade, Trilogy, or Actinovate.
Pythium	Seedlings rot off at soil line.	Water in seedlings with fish emulsion and seaweed at planting. Make sure beds are well drained. Apply Actinovate.
Rust	Reddish-brown lesions or pustules on both the upper and lower sides of the leaf.	Ensure good sanitation practices and good air circulation. Use drip irrigation. Remove infected leaves. Treat with *Bacillus subtilis* (Serenade).
Spider mite	Lighter leaf color. Tiny webbing on the underside of leaves.	Remove and destroy all infected leaves. Spray all sides of leaves with strong spray of water. Apply spinosad or garlic pepper tea.
Thrips	White blotches and streaking on leaves.	Release green lacewings and ladybeetles. Apply spinosad, neem, or insecticidal soap.

Insect and disease problems of lettuce

Insect/disease	Symptoms	Organic treatment
Alternaria leaf blight	Dark, irregular lesions on leaves.	Buy seeds from reputable sources. Apply Actinovate or Serenade.
Aster yellows	Stunted, yellow leaves.	Control leafhoppers. Remove infected plants. Keep weeds controlled in and around garden.
Cabbage looper	Gray-brown moth with "V" on forewings, active at night. Green "inchworms" (the larvae) eat ragged holes into leaves.	Cover plants with floating row covers at night. Apply Btk (*Bacillus thuringiensis* var. *kurstaki*) or spinosad.
Cutworm	Seedlings chewed through at soil level.	Apply Btk, spinosad, or diatomaceous earth. Release beneficial nematodes or trichogramma wasps.
Downy mildew	Pale yellow areas on upper leaf; white fuzzy growth below.	Remove infected leaves. Treat with Actinovate, Serenade, baking soda, or neem product.
Powdery mildew	White or gray "dust" covers plant.	Remove infected leaves. Treat with Actinovate, Serenade, baking soda, or neem product.

Insect and disease problems of nightshade family

Insect/disease	Symptoms	Organic treatment
Anthracnose	Small, black, circular lesions on ripe fruit.	Provide good air circulation, and remove infected fruit. Apply Serenade or neem products.
Aphid	Small, soft-bodied insect, usually on new growth, that can cause leaf discoloration and stunt new growth. Can also produce large quantities of a sticky exudate known as honeydew, which often turns black with the growth of a sooty mold fungus.	Release beneficial insects such as ladybeetles or green lacewings. Apply insecticidal soap, neem, or orange oil.
Bacterial leaf spot	Small, dark spots on the leaves, as well as larger, more angular sections of the leaf that appear scorched. Occurs in warmer temperatures.	Keep beds free of debris. Remove infected fruit and leaves. Apply wettable sulfur preemptively. Apply Actinovate or Serenade.
Bacterial speck	Tomato disease. Small, dark spots on the leaves, as well as larger, more angular sections of the leaf that appear scorched. Occurs in cooler temperatures.	Treat seeds in hot-water bath (122°F for 25 minutes). Keep beds free of debris and remove weeds from area. Rotate crops. Apply Serenade.
Blossom end rot	Blossom ends of the tomatoes (and the sides or ends of peppers) will be dark brown or black and sunken.	Remove and compost infected fruit. Test soil for calcium deficiency. Do not allow plants to dry out. Mulch plants.
Colorado potato beetle	Orange beetle with a yellow and black-striped shell. Both larvae and adults feed on the foliage and, if left untreated, can completely defoliate plants.	Rotate crops. Use floating row covers. Release beneficial insect predators. Apply pyrethrum products as last resort.
Downy mildew	Rust and yellow mottling on upper side of leaves. Fuzzy, gray spots on underside. Generally a late-season disease.	Remove infected leaves and destroy. Apply Serenade, Trilogy, or Actinovate.
Early blight	Lower leaves have dark spots and start to yellow.	Remove infected leaves. Apply Actinovate or Serenade.
Fusarium and verticillium wilt	Initial symptoms appear as isolated areas of yellowing and wilting. Entire or portions of leaves turn yellow. Wilting on lower or upper part of the plant, especially during heat of the day. Dark discolorations along stems.	Remove and destroy infected plant material. Plant resistant varieties. Apply Actinovate.
Gray leaf spot	Small, dark spots on the underside of the leaves that grow and take on a grayish appearance. Centers crack and fall out.	Remove and destroy infected plant material. Rotate crops.
Growth cracks	Concentric rings or fissures on the stem ends of fruit. Not a disease but a symptom of uneven watering.	Water consistently and evenly.
Late blight	More common in cooler temperatures. Problem with potato tubers that overwinter in soil. Appears as water-soaked spots on the margins of lower leaves that turn brown to purplish-black.	Buy tubers from reputable sources. Plant in well-drained soils. Keep gardens and surrounding areas free of weeds. Apply Serenade.
Leaf miner	White, winding tunneling inside leaves.	Mostly cosmetic. Remove and destroy infected leaves. Apply spinosad.

Nematodes, root-knot	Galls on roots. Wilted, stressed plants are lighter and appear to need water.	Add well-decomposed organic matter to soil before planting. Plant cover crops in the off season. Rotate crops. Apply compost tea or ClandoSan.
Powdery mildew	Light, blotchy gray spots on leaves. Most common in cooler, rainy months.	Remove infected leaves and destroy. Apply Serenade, Trilogy, or Actinovate.
Pythium	Seedlings rot off at soil line.	Water in seedlings with fish emulsion and seaweed at planting. Make sure beds are well drained. Apply Actinovate.
Scab of potato	Warty or sunken lesions on tubers that provide entry for other pathogens.	Rotate crops. Inspect tubers before planting for symptoms.
Septoria leaf spot	Brown lesions with yellow halos on lower leaves, especially on tomatoes.	Remove infected leaves. Keep beds free of debris. Use drip irrigation. Mulch. Apply Serenade.
Southern blight	White fungal growth that girdles and weakens stems.	Plant in well-drained soil. Rotate crops. Destroy infected plant debris.
Spider mite	Lighter leaf color. Tiny webbing on the underside of leaves.	Remove and destroy all infected leaves. Apply strong spray of water on all sides of leaves. Apply spinosad or garlic pepper tea.
Stem and stolen canker of potato	Dirtlike specs on tubers that will not wash off. Aboveground growth stunted and may be girdled at soil line.	Rotate crops. Inspect tubers before planting for symptoms.
Stink bug and leaf-footed bug	Shield-shaped insects with dark brown or green bodies suck juices from fruit and leave hard, discolored blisters.	Check for insects and check leaves for clusters of barrel-shaped, yellow to green eggs. Remove and destroy. Keep beds free of debris. Apply kaolin clay as preventive. Apply pyrethrum as last resort.
Sunscald	Not a disease, only a symptom. Tan or translucent scorched areas on fruit. Common on peppers and tomatoes in the heat of summer.	Furnish afternoon shade in hotter months.
Thrips	White blotches and streaking on leaves.	Release green lacewings or ladybeetles. Apply spinosad, neem, or insecticidal soap.
Tobacco and tomato hornworm	Large green or brown caterpillars of the hawk moths with "V" shape on their backs and horn shape on their tails. Can defoliate plants in a very short time.	Inspect and handpick from plants. Apply Bt (*Bacillus thuringiensis*). Release braconid wasps.
Tobacco mosaic virus	Mottling and deformation of leaves, common to viral diseases.	Remove and destroy infected plants. Rotate crops. Wash hands well after smoking before handling plant. Control insect vectors.

Insect and disease problems of okra

Insect/disease	Symptoms	Organic treatment
Aphid	Small, soft-bodied insect, usually on new growth, that can cause leaf discoloration and stunt new growth. Can also produce large quantities of a sticky exudate known as honeydew, which often turns black with the growth of a sooty mold fungus.	Release beneficial insects such as ladybeetles and green lacewings. Apply insecticidal soap, neem, or orange oil.
Flea beetle	Small beetles jump when leaves are brushed and bore tiny holes in leaves.	Keep debris cleaned from area. Apply garlic tea, neem, or pyrethrum products (as last resort).
Fusarium wilt	Yellowing of leaves that spreads to entire plant.	Remove and destroy infected plants. Apply Actinovate as a soil drench.
Nematodes, root-knot	Wilted, stressed plants are lighter and appear to need water.	Add well-decomposed organic matter to soil before planting. Plant cover crops in the off season. Rotate crops. Apply compost tea or ClandoSan.
Stink bug	Insect with shield-shaped, dark brown or green body. Barrel-shaped yellow to green eggs turn pink or gray on the underside of leaves. Blisters on fruit.	Inspect underside of leaves daily. Remove both eggs and adult insects. Keep debris cleaned from beds.
Thrips	White blotches and streaking on leaves.	Release green lacewings and ladybeetles. Apply spinosad, neem, or insecticidal soap.

Insect and disease problems of onions

Insect/disease	Symptoms	Organic treatment
Botrytis leaf blight	Brownish "dust" on plant leaves, especially in wet, humid weather.	Remove infected leaves and debris. Apply Actinovate, Serenade, or neem products.
Pink root	Roots turn pink, and bulbs shrivel.	Plant resistant varieties.
Purple blotch	Sunken, purple lesions on leaves.	Apply Actinovate or Serenade.
Thrips	White blotches and streaking on leaves.	Release green lacewings and ladybeetles. Apply spinosad, neem oil, or insecticidal soap.

Insect and disease problems of sweet potatoes

Insect/disease	Symptoms	Organic treatment
Black rot	Dark brown or black, circular, depressed areas on tubers. Blackened areas and decay at base of stem.	Purchase tubers from reputable nurseries. Dig tubers carefully. Remove infected plants and destroy. Do not replant in area for a few years.
Downy mildew	Rust and yellow mottling on upper side of leaves. Fuzzy, gray spots on underside. Generally a late-season disease.	Remove infected leaves and destroy. Apply Serenade, Trilogy, or Actinovate.
Nematodes, root-knot	Wilted, stressed plants are lighter and appear to need water.	Add well-decomposed organic matter to soil before planting. Plant cover crops in the off season. Rotate crops. Apply compost tea or ClandoSan.
Powdery mildew	Light, blotchy gray spots on leaves. Most common in cooler, rainy months.	Remove infected leaves and destroy. Apply Serenade, Trilogy, or Actinovate.
Stem rot	Young plants turn yellow; inside of stems turn brown or black. Small tubers with decay on stem end.	Purchase disease-free slips from reputable nurseries.
Sweet potato flea beetle	Small, black beetles jump when leaves are brushed and bore tiny holes in leaves. Larvae feed on roots.	Keep debris cleaned from area. Treat with garlic tea, neem, or pyrethrum products (as last resort).
Sweet potato weevil	Larvae of beetle with red head and long red snout and shiny, black body burrow into tubers, causing them to be spongy. Yellow, wilted foliage.	Remove and destroy entire plant. Rotate crops. Release parasitic wasps and beneficial nematodes (*Heterorhabditis bacteriophora*).
Wireworm	Shiny, reddish-brown to yellow, wormlike insect that feeds underground and burrows into tubers.	Turn soil well before planting to expose larvae. Rotate crop locations. Plant cover crops in the off season. Release beneficial nematodes. Apply wood ashes. Destroy infected tubers.

☀ Appendix C
Seed and Plant Sources

The nurseries listed in this section are in no way meant to exclude any nursery or other business that sells edible plants and seeds. Local retail nurseries are always good sources for all edible gardening needs and usually carry products specific to the area. Big chain stores may be the exception to this rule.

Trees, Bramble Fruits, and Grapes

Bob Wells Nursery
17160 CR 4100
Lindale, TX 75771
903-882-3550
Fruit trees, bramble fruits, and grapes

Brazos Citrus Nursery
P.O. Box 167
West Columbia, TX 77486
979-345-2906
Wholesale fruit trees

Fanick's Garden Center, Inc.
1025 Holmgreen Road
San Antonio, TX 78220

210-648-1303
Pecan, citrus, olive, and fruit trees; bramble fruits; grapes; and rootstock

Peerless Farm
3176 FR 2400
Bigfoot, TX 78005
830-663-3651
Wholesale fruit trees

R & T Quality Nursery
P.O. Box 733
19122 Hwy 69N
Lindale, TX 75771
903-881-0600
Pecan and fruit trees, blueberry shrubs, bramble fruits, and grapes

Rio Grande Nursery
4195 North Bryan Road
Mission, TX 78573
956-581-4880
Citrus trees

Texas Pecan Nursery, Inc.
P.O. Box 306
Chandler, TX 75758
903-849-6203
Pecan trees

Womack Nursery
2551 State Hwy 6
De Leon, TX 76444
254-893-6497
Pecan and fruit trees, asparagus, bramble fruits, grapes, and graftwood

Hybrid and Heirloom Seeds, Bedding Plants, Potato Sets, Garlic, Asparagus Crowns, and Strawberry Plants

Baker Creek Heirloom Seed Co.
2278 Baker Creek Road
Mansfield, MO 65704
417-924-8917

Bountiful Gardens
1726-D South Main
Willits, CA 95490
707-459-6410

D. Landreth Seed Company
60 East High Street, Bldg #4
New Freedom, PA 17349
1-800-654-2407

Gabriel Valley Farm
440 SH 29
Georgetown, TX 78626
512-930-0923
Wholesale only, organic vegetable and herbs, bedding plants

Gourmet Garlic Gardens
325-348-3049
Garlic, online orders only: http://www.gourmet garlicgardens.com/

Johnny's Selected Seeds
955 Benton Avenue
Winslow, ME 04901
207-238-5327

Nichols Garden Nursery
1190 Old Salem Road NE
Albany, OR 97321
800-422-3985

Seed Savers Exchange
3094 North Winn Road
Decorah, IA 52101
563-382-5990

Seeds of Change
P.O. Box 4908
Rancho Dominguez, CA 90220
888-762-7333

Index

Abelmoschus esculentus (okra), 223–26, 240, 264

access considerations. *See* pathways

Aci Sivri variety (pepper), 216

Actinovate, fungal diseases, 242

Adam's variety (elderberry), 129

Adirondack Red variety (potato), 219

Advance variety (loquat), 104–105

aeration, compost heaps, 21–22

Agate variety (soybeans), 205

agricultural growing regions, for site evaluation, 9

Aji Amarillo variety (pepper), 216

Alabama Red variety (okra), 225

Alapaha variety (blueberry), 127

Albino variety (beets), 166

Alderman variety (English pea), 206

alfalfa meal, 23

Ali Baba variety (watermelon), 194

Allium schoenoprasum (chives), 153

Allium spp. (onion). *See* onion family

Allred variety (plum), 123

Allstar variety (strawberry), 149

Alma variety (fig shrub), 131

amaranth, 4*f*, 162–65, 253

Amber Globe variety (turnip), 187

Ambrosia variety (corn), 174

amendments, soil, 9, 23–25
   See also compost heaps; fertilizer guidelines, overviews

American Flag variety (leek), 229

Americano Tonda variety (pumpkin), 198

American variety (persimmon), 113

Amity variety (raspberry), 140

Anaheim Chile variety (pepper), 216

Ananas d'Amerique a Chair Verte variety (muskmelon), 192–93

Ancho variety (pepper), 216

Anderson, Bob, 228

Anethum graveolens (dill), 154

Angel Red variety (pomegranate), 115

anise, 150

Anna variety (apple), 92

Anthriscus cerefolium (chervil), 153

Apache variety (blackberry), 139

apartment gardens, 15

aphids. *See* insect pests, overviews

Apollo variety (asparagus), 137

Appaloosa variety (bush bean), 204

apple trees, 18, 89–93, 247

apricot trees, 116–20, 247

Arapaho variety (blackberry), 139

Arbequina variety (olive), 84

arbor structures, 16*f*, 18

Arbosana variety (olive), 84

arching plants, listed, 40

architectural/structural elements
   design considerations, 12–13
   Japanese gardens, 62
   pizza gardens, 64

for vertical space, 17–18
   walled courtyard gardens, 66, 68

Arkansas Black variety (apple), 92

Armking variety (nectarine), 121

artichoke plants, 133–35

Ashworth variety (corn), 174

Asian pear varieties, 110–11

Asian Tempest variety (garlic), 228

Asmara variety (soybeans), 205

asparagus plants, 135–37, 250

Aster Café, Lake Austin Spa Resort, 46–50

Atlas variety (asparagus), 137

Aurea variety (elderberry), 129

Austin variety (blueberry), 127

Austin variety (southern dewberry), 141

Australian Butter variety (pumpkin), 198

Australian Yellow variety (lettuce), 209

Autumn Bliss variety (raspberry), 140

Ayers variety (pear), 110

B9 rootstock (apple), 91

Bababerry variety (raspberry), 140

Baby Bear variety (pumpkin), 198

Baby Corn variety (corn), 174

Bacalan de Rennes variety (cabbage), 180

Bacillus thuringiensis (Bt), insect pests, 243

backflow devices, 30

baking soda, fungal diseases, 242

balcony gardens, 15

Banana variety (muskmelon), 193

bark mulch, 33

Barouni variety (olive), 84

Bartholomew, Mel, 73

basil, 150–51

bat guano, 23

bat houses, 87, 187

bay, sweet, 151

bean plants

    children's gardens, 51, 54f

    cultivation guidelines, 18, 199, 201–202

    pests summarized, 260

    varieties, 200f, 202–205, 239

Beaucamp, Stephane, 46, 59

Beauregard variety (sweet potato), 234

Beauveria bassiana, insect pests, 243

Beck's Big Buck variety (okra), 225

bed access. *See* pathways

beds, raised, 25–26, 59

bee balm, 152

bee boxes, 61–62

Beer Friend variety (soybeans), 205

beet family

    overview, 162–63

    amaranth, 163–65

    beets, 36f, 165–66

    pests summarized, 253

    spinach, 166–67, 241

    Swiss chard, 36f, 168–69, 239

Belgian White variety (carrot), 170

Bella Vista Olive Orchard, 81f, 85

Belle of Georgia variety (peach), 121

beneficial insects, 87–88, 243–44

Bennings Green Tint Scallop variety (summer squash), 196

Beta vulgaris spp. *See* beet family

Betty's variety (sweet potato), 234

Big Jim variety (loquat), 105

Big Max variety (pumpkin), 198

Big Red Ripper variety (southern pea), 208

bilateral cordon system, pruning, 143

biodynamic farming, 127–28

bird pests, 126, 127, 143–44

birds, children's gardens, 52

Black Beauty variety (eggplant), 214

Black Beauty variety (elderberry), 129

Black Beauty variety (mulberry), 106

Black Beauty variety (Muscadine grape), 146

blackberry vines, 18, 137–39

Black Diamond variety (watermelon), 194

black-eyed peas, 207–208

Black Fatsu variety (winter squash), 197

Black Lace variety (elderberry), 129

Blackland Prairie soils, 9

Black mulberry species, 106

Black Pearl variety (pepper), 216

Black Seeded Simpson variety (lettuce), 209

Black Spanish varieties (radish), 186

Black Tail Mountain variety (watermelon), 194

Black Turtle variety (bush bean), 204

Blanc du Bois variety (grape), 144

Blauwschokker variety (English pea), 206

Blenheim variety (apricot), 119

blight. *See* diseases, overviews

blood meal, 23

Blood Orange variety (orange), 100

Bloomsdale varieties (spinach), 167

blueberry shrubs, 125–28

Blue Coco variety (pole bean), 202

Blue Hubbard variety (winter squash), 197

Blue Lake 274 variety (bush bean), 204

Blue Lake Stringless variety (pole bean), 202

blue plants, listed, 41

Blush variety (eggplant), 214

Bok Choy variety (cabbage), 180

bold-textured plants, 38

Bolita variety (bush bean), 204

bone meal, 23, 118

Bordeaux mixture, fungal diseases, 242

border plants, listed, 39

borders, parterre gardens, 45

boron deficiency, beets, 166

Boston Marrow variety (pumpkin), 198

Boston Pickling variety (cucumber), 191

Boule d'Or variety (muskmelon), 193

Bountiful variety (bush bean), 204

Bounty variety (peach), 121

Bowling Red variety (okra), 225

boxes, square foot gardening, 73

boxwood varieties, 45

Braeburn variety (apple), 92

bramble fruits, 137–41, 251

Brassica spp. *See* crucifers

Brazos variety (blackberry), 139

Brightwell variety (blueberry), 127

Brinkley White variety (sweet potato), 234

Brison variety (blackberry), 139

broccoli, 175–78, 239, 256–57

Broccoli Raab variety, 177

Broccoverde variety (cauliflower), 182

Bronze Arrow variety (lettuce), 209

Brown Crowder variety (southern pea), 208

Brown Turkey variety (fig shrub), 131

Brunswick variety (cabbage), 180

Brussels sprouts, 175–76, 178–79, 239, 256–57

Bryan variety (apricot), 119

Buhl variety (corn), 174

Bull's Blood variety (beet), 166

Burgundy variety (okra), 225

Burmese variety (okra), 225

burnet, salad, 152

Burpee Stringless variety (bush bean), 204

bush beans, 201–202, 204–205, 239, 260

Bush Kentucky Wonder variety (bush bean), 204

Bush Puerto Rico variety (sweet potato), 234

Butterbeans variety (soybean), 205

Buttercup variety (winter squash), 197

butterfly gardens, 52

Byrd Pecan Orchard, 85*f*, 88–89

cabbages, 161*f*, 175–76, 179–81, 239, 256–57

Caddo variety (pecan), 88

Calabrese variety (broccoli), 177

Calico Crowder variety (southern pea), 208

California Mission variety (olive), 84

California varieties (garlic), 228

California Wonder variety (pepper), 216

cane system, pruning, 143

cantaloupe varieties, 239

caprifig, 130

Capsicum annum (pepper), 211–13, 215–17

Cardinal variety (strawberry), 149

Caribe variety (potato), 220

Carlos variety (Muscadine grape), 146

Carola variety (potato), 220

Caroline Everbearing variety (raspberry), 140

Carouby de Mausanne variety (Snow pea), 207

carrots, 169–71, 239, 254

Carya illinoinensis (pecan), 85–89, 246

Casaba Golden Beauty variety (muskmelon), 194

Cascadia variety (sugar snap pea), 207

Catskill variety (Brussels sprouts), 179

cauliflower, 175–76, 181–82, 239, 256–57

Cayenne variety (pepper), 216

Celeste variety (fig shrub), 132

Centennial variety (sweet potato), 234

chamomile, Roman, 152–53

Champagne variety (loquat), 105

Champanel variety (grape), 144

Chandler variety (strawberry), 149

Changsha variety (tangerine), 101

Chantenay Red Core variety (carrot), 170

Charantais melon variety, 192*f*, 193

chard, Swiss, 36*f*, 168–69, 239, 253

Charentais variety (muskmelon), 193

Charleston Gray variety (watermelon), 194

Cherokee Purple variety (tomato), 223

Cherokee Trail of Tears variety (pole bean), 202

Cherry varieties (tomato), 241

chervil, 153

Cheynne Bush variety (pumpkin), 198

Chicago Pickling variety (cucumber), 191

Chicamaw Farm, 127–28

Chico variety (jujube), 101–102

Chieftain Savoy variety (cabbage), 180

children's gardens, 50–56, 201

chilling hours, site evaluation, 11

Chinese cabbage, 161*f*, 179, 180–81, 239

Chinese variety (apricot), 120

Chinese Yellow variety (cucumber), 190

Chiogga variety (beet), 166

chives, 45, 153

chlorosis, iron, 28, 143

Chocolate variety (persimmon), 113

Choctaw variety (blackberry), 139

Chojuro variety (pear), 110

Choppee variety (okra), 225

Christmas Red Calico variety (pole bean), 202

cilantro, 153–54

Cimarron Red variety (Romaine lettuce), 210

Cisneros Grande variety (tomatillo), 221

Citrullus lanatus (watermelon), 191–92, 194–95

citrus oil, insect pests, 244

citrus trees, 77–80, 93–101, 248

city regulations, garden design, 57, 60

clay soils, 5, 9, 19, 21

Clementine variety (tangerine), 101

Clemson Spineless variety (okra), 225

Climax variety (blueberry), 127

clover cover crops, 31

cocoa shell mulch, 33

Cocozelle Bush variety (summer squash), 196

cold frames, children's gardens, 52

collard greens, 182

Collier variety (mulberry), 106–107

color element, in designing gardens, 40–43

Colossus Crowder variety (southern pea), 208

columnar plants, listed, 40

community gardens, 15

community regulations, site evaluation, 7

compost applications, overview, 27

   See also specific plants, e.g., asparagus; blackberry vines

compost heaps, 21–23, 46, 52

compost tea, 27

Congo variety (watermelon), 194

container gardens, 15

Contender variety (bush bean), 204

Contorta variety (jujube), 103
Coratina variety (olive), 84
coriander, 153–54
Coriandrum sativum (cilantro), 153–54
corn, 171–74, 239–40, 255
Cornell's Bush Delicata variety (winter
    squash), 197
corn gluten meal, 23
Cosmic Purple variety (carrot), 170
Cosmo variety (Romaine lettuce), 210
cotton root rot. *See* diseases, overviews
cottonseed meal, 24
Couer de Boeuf variety (cabbage), 180
Country Gentleman variety (corn), 174
courtyard gardens, 66, 68–70
cover crops, 31
Cow Horn variety (okra), 225
Cranberry Red variety (potato), 220
Crane variety (muskmelon), 193
Crapaudine variety (beet), 166
Crenshaw variety (muskmelon), 193
Creole variety (garlic), 228
Crimson Gold variety (nectarine), 121
Crimson Sweet variety (watermelon),
    194
crucifers
        overview, 175–76
        broccoli, 176–78, 239
        Brussels sprouts, 178–79, 239
        cabbages, 161*f*, 179–81, 239
        cauliflower, 181–82, 239
        mustard greens, 182, 183–84,
            240
        pests summarized, 256–57
        radishes, 185–86, 241
        turnips, 186–87, 241
Cubanelle variety (pepper), 216
cucumbers, 18, 187–91, 240, 258–59
Cucumis spp. *See* cucurbits
cucurbits

overview, 18, 187–89
cucumbers, 189–91, 241
melons, 191–95, 239, 241
pests summarized, 258–59
squashes, 195–99, 240, 241
Cumberland Black variety (raspberry),
    140
Cushaw Green Striped Crookneck
    variety (pumpkin), 198
Cut Leaf variety (elderberry), 129
Cylindra variety (beet), 166
Cymbopogon citratus (lemongrass),
    155–56
Cynara scolymus (artichoke), 133–35
Cynthiana variety (grape), 144–45

daikon radishes, 185–86
Daisy variety (watermelon), 194
damping off disease, 160, 209, 212
    See also diseases, overviews
Dancy variety (tangerine), 101
Danish Ballhead variety (cabbage), 180
Danvers varieties (carrot), 170
Danyelle variety (lettuce), 210
Dark Red Norland variety (potato), 220
Darlene variety (Muscadine grape), 146
Daucus carota (carrot), 169–71
Dawn Giant variety (leek), 229
Dean's Purple variety (pole bean), 202
De Bourbonne variety (cucumber), 191
deer fencing, 16–17
Deer Tongue variety (lettuce), 210
De Grace variety (Snow pea), 207
Delicata variety (winter squash), 197
De Milpa variety (tomatillo), 221
Denman variety (peach), 121
dent corn, 171
Desert King variety (watermelon), 194
designing gardens, considerations
        layout arrangements, 12–15

plant elements, 37–42
structural elements, 16–18
time requirements, 13, 15
designing gardens, style approaches
    children's spaces, 50–56
    commercial spaces, 59–62
    Japanese, 62–65
    parterre, 43–45, 46–50
    pizza gardens, 66, 67*f*
    potager/kitchen, 43, 45–46
    residential spaces, 56–59
    square foot approach, 73
    walled courtyards, 66, 68–70
Desirable variety (pecan), 88
Detroit Dark Red variety (beets), 166
dewberry, southern, 141
diatomaceous earth, insect pests,
    244
Di Cicco variety (broccoli), 177
dill, 154
diseases, overviews
    fungal treatments, 242–43
    perennial plants, 250–52
    trees, 246–50
    vegetables, 199, 250, 252–65
    See also specific plants, e.g.,
        grapevines; tomatoes
Dixieland variety (peach), 121
Dixie Queen variety (watermelon), 194
Dixie Speckled Butter Peas variety (bush
    bean), 204
Doc Hill Golden Bell variety (pepper),
    216
dogs, 33
Dorman Red variety (raspberry), 140
Dorsett Golden variety (apple), 92
Dougherty, Jack, 85
drainage patterns, 6, 54, 57
drip irrigation, 29–30
Drunken Woman variety (lettuce), 210

Dutch Yellow variety (shallot), 232
Dwarf Blue Scotch variety (kale), 183

Eagle Pass variety (okra), 226
Earligrande variety (peach), 122
Early Flat Dutch variety (cabbage), 180
Early Fry variety (Muscadine grape), 146
Early Golden variety (apricot), 120
Early Hanover variety (muskmelon), 193
Early Jersey Wakefield variety (cabbage), 180
Early Prolific Yellow Straightneck variety (summer squash), 196
Early Red variety (loquat), 105
Early Snowball variety (cauliflower), 182
Early White Bush Scallop variety (summer squash), 196–97
Early White Vienna variety (kohlrabi), 185
earthworms, 24
Eden's Gem variety (muskmelon), 193
Edisto 47 variety (muskmelon), 193
Edmonson variety (cucumber), 190
Edna Evans variety (sweet potato), 234
Edwards Plateau soils, 9
eggplant, 211–14, 240, 262–63
Ein Shemer variety (apple), 92
Elberta variety (peach), 122
elderberry shrubs, 128–30
Elephant Head variety (amaranth), 164
Elephant variety (garlic), 228
El Paso, rainfall, 10
Emerald variety (okra), 226
Empire variety (apple), 92
English pea varieties, 206–207
Eriobotrya japonica (loquat), 103–105
espalier training, 40, 69–70, 79
Eureka variety (lemon), 97–98
Eureka variety (persimmon), 113
Eva Purple Ball variety (tomato), 223

Even Star Land Race variety (collard greens), 182
ever-bearing strawberries, 147
evergreen herbs, parterre gardens, 45
evergreen plants, listed, 41
Everhard variety (orange), 100
Everona Large Green variety (tomatillo), 221
Excel variety (sweet potato), 234
expanded shale, 24

Fairmont Dallas hotel, 60f, 61–62
Falstaff variety (Brussels sprouts), 179
feather meal, 24
fencing, 7, 16–18, 59
fennel, 154, 155f
fertilizer guidelines, overviews
        nutrients for, 23–25, 27–28
        perennial plants, 133
        trees, 95, 117–18
        vegetables, 176, 189, 199, 201, 213
        See also specific plants, e.g., blueberry shrubs; corn; pecan trees
fig shrubs, 18, 52, 56, 130–32, 250
Fin de Meaux variety (cucumber), 191
fine-textured plants, 38
fire blight. *See* diseases, overviews
fish emulsion and seaweed, 27, 28
fish meal, 24
Flamingo Pink variety (Swiss chard), 168
Flashy Butter Oak variety (lettuce), 210
Flat of Egypt variety (beets), 166
Floridaking variety (peach), 122
Florida Speckled Butter variety (pole bean), 202–203
Foeniculum vulgare (fennel), 154, 155f
foliar applications, 27
Fordhook Acorn variety (winter squash), 197

Fordhook Giant variety (Swiss chard), 168
Forellen-Schluss variety (Romaine lettuce), 210–11
Forkert variety (pecan), 88
form element, in designing gardens, 40
four-year plan, 13, 15
Fragaria spp. (strawberry), 147–49, 252
Frantonio variety (olive), 85
Frazier White variety (sweet potato), 234
freeze protection. *See specific plants, e.g.,* persimmon trees
French Fingerling variety (potato), 220
French Flageolet variety (bush bean), 204
French Horticultural variety (pole bean), 203
French Red variety (shallot), 232
French variety (Swiss chard), 168
fruit trees
        apple, 18, 89–93
        citrus types, 93–116
        parterre gardens, 45
        pear, 12f, 107–11
        pests summarized, 247–248249
        stone fruit types, 116–24
fruit types, in designing gardens, 13
Fuji variety (apple), 92
fungal diseases
        damping off, 160, 209, 212
        inoculations against, 173
        irrigation cautions, 29
        organic treatments, 242
        See also diseases, overviews
Fuyu variety (persimmon), 113

Gai Lohn variety (cabbage), 181
Gala variety (apple), 92
Galeux d'Eysines variety (pumpkin), 198
garlic, 36f, 227–28, 240, 264

Gaucho variety (muskmelon), 193

Georgia Southern variety (collard greens), 182

Geraldi Dwarf variety (mulberry), 107

germander varieties, 45

Giant Musselburg variety (leek), 229

Giant Noble variety (spinach), 167

Giant of Naples variety (cauliflower), 182

Giant Winter variety (spinach), 167

Ginseng Red variety (sweet potato), 234

globe basil, 151

Glycine max (soybeans), 205

Goldbeere variety (elderberry), 129

Golden Bantam variety (corn), 174

Golden Delicious variety (apple), 92

Golden Giant variety (amaranth), 164

Golden Midget variety (watermelon), 194

Golden Slipper variety (sweet potato), 234

Golden Sunrise variety (Swiss chard), 168

Golden variety (beets), 166

Gold-Kist variety (apricot), 120

Gold Nugget variety (loquat), 105

Goyo Kumbo variety (eggplant), 214

Granada variety (pomegranate), 115

Grande variety (asparagus), 137

granite dust/mulch, 24, 31, 33

Granny Smith variety (apple), 92

grapefruit trees, 77–80, 95–96, 248

grapevines, 18, 45, 141–46, 252

Gray Grisell variety (shallot), 232

gray plants, listed, 42

gray water, 72

Great Wall variety (persimmon), 113

Greek Giant variety (amaranth), 164

Green Apple variety (eggplant), 214

Green Arrow variety (English pea), 206

green cabbage varieties, 180

Green Callaloo variety (amaranth), 164

Green Gate Farms, 227f

Green Glaze variety (collard greens), 182

Green Globe varieties (artichoke), 135

Green Macerata variety (cauliflower), 182

Green Machine variety (muskmelon), 193

Green Nutmeg variety (muskmelon), 193

greens (*Brassica* spp.), 182–84, 240, 256–57

Green Wave variety (mustard green), 183

Green Zebra variety (tomato), 223

ground cover plants, listed, 39

guano, bat, 23

Gulfgold variety (plum), 124

Gulf Ruby variety (plum), 123

Gwang Yang variety (persimmon), 113

gypsum, 24

Habanero variety (pepper), 216

Hachiya variety (persimmon), 113

Hale's Best variety (muskmelon), 193

Halford rootstock, 116–17

hardiness zones, site evaluation, 8

Harglow variety (apricot), 120

Hartmann's Giant variety (amaranth), 164

Harvester variety (peach), 122

harvest guidelines. *See specific plants, e.g.,* blueberry shrubs; persimmon trees

Haschberg variety (elderberry), 129

haydite, 24

Hayman variety (sweet potato), 234

hay mulch, 33

Healthmaster variety (carrot), 170

Hearts of Gold variety (muskmelon), 193

Helda-Romano variety (pole bean), 203

Helianthus tuberosus (Jerusalem artichoke), 146–47

Henderson/Ray variety (grapefruit), 95

Henderson's Black Valentine variety (pole bean), 203

Henderson's Charleston Wakefield variety (cabbage), 180

Henry, Jim, 82

Herbemont variety (grape), 145

herbicides, 33

herbs, 45, 61, 150–60

Heritage Red variety (raspberry), 140

Hernandez variety (sweet potato), 235

highbush varieties (blueberry), 127

Hill Country Heirloom Red variety (okra), 226

Hmong Red variety (cucumber), 190

HOA regulations, site evaluation, 7

Holland variety (apple), 92

homeowner regulations, site evaluation, 7

Honan Red variety (persimmon), 113

honey bees, 61–62

Honeydew Green variety (muskmelon), 193–94

Honey Jar variety (jujube), 103

Honey Rock variety (muskmelon), 193

Hong Kong species (kumquat), 97

horticulture oil, insect pests, 244

Hosui variety (pear), 110

Hot Banana Pepper variety, 216

hot varieties (pepper), 216, 240

Houston, rainfall, 10

Hubei variety (amaranth), 164

humus, 21

Ideal Purple Top Milan variety (turnip), 187

Illini Star variety (tomato), 223

Illinois Everbearing variety (mulberry), 107

Imperator variety (carrot), 170

Improved Meyer variety (lemon), 98
Indiana Gold variety (sweet potato), 235
Indian Cling variety (peach), 122
insecticidal soap, insect pests, 244
insect pests, overviews
    organic treatments, 243–45
    perennial plants, 250–52
    trees, 246–50
    vegetables, 250, 252–65
insects
    children's gardens, 52
    compost heaps, 22–23
Ipomoea batatas (sweet potato), 232–35
iron chelates, 28
irrigation guidelines, 29–30
Ison variety (Muscadine grape), 146
Italian stone pine, 77–81
Ivis White Cream variety (sweet potato), 235
Izu variety (persimmon), 113

Jack Be Little variety (pumpkin), 198
Jade variety (okra), 226
Jalapeño variety (pepper), 216
Japanese Climbing variety (cucumber), 190
Japanese gardens, 62–64
Japanese Long variety (cucumber), 189*f,* 190
Jarrahdale variety (pumpkin), 198
Jenny Lind variety (muskmelon), 194
Jersey Mac variety (apple), 92
Jersey varieties (asparagus), 137
Jerusalem artichokes, 146–47
Jewel variety (sweet potato), 235
Jimmy Nardello variety (pepper), 216–17
Jimmy T's variety (okra), 226
Jiro variety (persimmon), 113
John's variety (elderberry), 129
Jonagold variety (apple), 92

Jubilee variety (watermelon), 194
Juglans microcarpa (Texas walnut tree), 89
jujube trees, 77–80, 101–103, 248
June-bearing strawberries, 147
June Gold variety (peach), 122
Juneprince variety (peach), 122

Kaffir variety (lime), 98
kale, 175–76, 182, 183, 240, 256–57
Kanza variety (pecan), 88, 89
Katy variety (apricot), 120
kelp meal, 24
Kennebec variety (potato), 220
Kentucky Wonder varieties (pole bean), 203
Key Lime variety, 98
Kieffer variety (pear), 110
King of Mammoth variety (pumpkin), 198
King of the Garden variety (pole bean), 203
King Richard variety (leek), 229
Kiowa variety (blackberry), 139
kitchen gardens, designing, 43, 45–46, 59–62
knot gardens, 45
kohlrabi, 175–76, 184–85, 256–57
Komatsuna variety (cabbage), 181
Korean Purple variety (sweet potato), 235
kumquat trees, 77–80, 96–97, 248
Kyungsun Ban-Si variety (persimmon), 113

laboratories, soil testing, 25
lacewings, green, 243
Lactuca sativa (lettuce), 208–11, 240, 261
ladybeetles, 88, 243

La Feliciana variety (peach), 122
Lagenaria Longissima variety (summer squash), 197
Lake Austin Spa Resort, gardens, 46–50, 59
Lakota variety (pecan), 89
Lakota variety (pumpkin), 198–99
Landress Stringless variety (bush bean), 204
langbeinite, 24
Lang variety (jujube), 103
"lasagna beds," 26
Late Flat Dutch variety (cabbage), 180
Laurus nobilis (sweet bay), 151
layout arrangements. *See* designing gardens, considerations
Le Conte variety (pear), 110
leeks, 226–27, 228–29, 264
legumes
    beans, 200*f,* 201–205, 239
    for cover crops, 21, 31
    cultivation guidelines, 199, 201
    peas, 205–208, 240
    pests summarized, 260
lemon balm, 155
lemon basil, 151
lemongrass, 155–56
lemon trees, 77–80, 97–98, 248
Lemon variety (cucumber), 190
Le Noir variety (grape), 145
lettuce, 208–11, 240, 261
lima bean varieties, 202–203, 204, 205, 239
Limequat variety (lime), 98
limestone, agricultural, 23
lime trees, 77–80, 98, 248
Lincoln variety (English pea), 206
Lincoln variety (leek), 229
Lisbon variety (lemon), 97, 98
Listada de Gandia variety (eggplant), 214

Little Gem variety (Romaine lettuce), 211

Little Marvel variety (English pea), 206

Li variety (jujube), 103

Lollo varieties (lettuce), 210

Lomanto variety (grape), 145

Long Green Improved variety (cucumber), 190

Long Island Improved variety (Brussels sprouts), 179

Long Purple variety (eggplant), 214

loquat trees
    commercial gardens, 60–61
    cultivation guidelines, 18, 77–80, 103–104
    hardiness zones, 8
    pests summarized, 249
    varieties, 104–105

Loring variety (peach), 122

Louisiana Long Green variety (eggplant), 214

Love-Lies-Bleeding variety (amaranth), 165

Lovell rootstock, 116–17

Lucullus variety (Swiss chard), 168

Lunar White variety (carrot), 170

Lutz variety (beets), 166

Magness variety (pear), 110–11

Malabar variety (spinach), 167

Malali variety (watermelon), 194

Malus domestica (apple), 18, 89–93, 247

Mammoth Melting Sugar variety (Snow pea), 207

Mammoth Red Rock variety (cabbage), 180

Mandan variety (pecan), 89

mandarin orange trees, 77–80, 98–99

manures, compost heaps, 23

Marburger Orchard, 117*f*

Marconi variety (pepper), 217

Marglobe variety (tomato), 223

Marina di Chioggia variety (winter squash), 197

marjoram, 156

Marrs variety (orange), 100–101

Marumi species (kumquat), 97

Matt's Wild Cherry variety (tomato), 212*f*, 223

Maurino variety (olive), 85

Mayo Indian variety (amaranth), 165

mazes, children's gardens, 51

McBeth variety (loquat), 105

McCaslan variety (pole bean), 203

McCranie, Bill and Nancy, 127–28

McKinney, Eleanor, 46, 47–50

Mediterranean variety (mandarin orange), 99

Meiwa species (kumquat), 97

Melissa officinalis (lemon balm), 155

melons, 18, 187–89, 191–95, 239, 258–59

Melrose variety (pepper), 217

Mentha spp. (mint), 150*f*, 157

Merlo Nero variety (spinach), 167

Merlot variety (lettuce), 208*f*, 210

Methley variety (plum), 124

Mexican mint marigold, 156–57

Mexican Sour Gherkin variety (cucumber), 191

Meyer variety (lemon), 97, 98

Michihili Chinese variety (cabbage), 181

Mignonette Bronze variety (lettuce), 210

Miho variety (mandarin orange), 99

mildews. *See* diseases, overviews

Miniature Bell variety (pepper), 217

mint, 150*f*, 157

Miriah variety (amaranth), 165

Misato varieties (radish), 186

Mississippi Silver variety (southern pea), 208

Miss Kim variety (persimmon), 113

Missouri Wonder variety (pole bean), 203

MM rootstocks (apple), 91–92

molasses and seaweed, 27, 28

Mollie's Delicious variety (apple), 92

Molten Fire variety (amaranth), 165

Monarda didyma (bee balm), 152

Monroe variety (peach), 122

Moon and Stars variety (watermelon), 194

Moon Cake variety (soybeans), 205

Moonglow variety (pear), 111

Moorpark variety (apricot), 120

Morris Heading variety (collard greens), 182

Morris variety (plum), 124

Morus spp. (mulberry), 105–107

Mr. Big variety (English pea), 206

M rootstocks (apple), 91

mulberry plants, 52, 56, 105–107

mulches, 31–33, 126

Muscadine grapevines, 145–46

muskmelons, 191–94, 258–59

Musquee de Provence variety (pumpkin), 199

mustard greens, 175–76, 182, 183–84, 240

mycorrhizal fungi, 21, 89, 133
    See also *Rhizobium* bacteria

N-33 Navel variety (orange), 100–101

Nacono variety (pecan), 89

Nagami species (kumquat), 97

Nana Dwarf variety (pomegranate), 116

Nantes varieties (carrot), 170

Natera, Andre, 60*f*, 61

Navaho variety (blackberry), 139

Navet des Vertus Marteau variety (turnip), 187

nectarine trees, 116–18, 121
neem oil, 242, 245
neighborhood gardens, 15
neighbor relations, site evaluation, 7
Nemaguard rootstock, 116
nematodes, 116–17, 243
   See also diseases, overviews
New Red Fire variety (lettuce), 210
nightshade family
   overview, 211–13
   eggplant, 213–14, 240
   peppers, 214–17, 240
   pests summarized, 262–63
   potatoes, 217–20, 240
   tomatillos, 220–21
   tomatoes, 18, 221–23, 241
Nijisiki variety (pear), 111
nitrogen, 23–25, 199
Noir de Carmes variety (muskmelon), 193
Northpole Columnar variety (apple), 92
Nosema locustae, insect pests, 245
Nova variety (elderberry), 129
Nova variety (raspberry), 140
nurseries, listing of, 266–68

Oakleaf variety (lettuce), 210
Ochlockonee variety (blueberry), 127
Ocimum basilicum (basil), 150–51
Oconee variety (pecan), 88
OH rootstocks (pear), 109
okra, 223–26, 240, 264
Olea europeaea (olive), 81–85
Olive Tree Learning Center, 54–56
olive trees, 77–85
onion family
   overview, 226–27
   garlic, 227–28
   leeks, 228–29
   onions, 229–31, 240

   pests summarized, 264
   shallots, 231–32
Orange Giant variety (amaranth), 165
Orangeglo variety (watermelon), 194
orange plants, listed, 41
orange trees, 93*f*, 99–101
oregano, 157–58
Oregon Giant variety (Snow pea), 207
Oregon Sugar Pod II variety (Snow pea), 207
Oregon Trail variety (English pea), 207
Oriental Giant variety (spinach), 167
Orient variety (pear), 111
Origanum spp., 156–58
Oriole Orange variety (Swiss chard), 168
Orlando Seedless variety (grape), 145
ornamental trees, 39, 89–93
Oro Blanco variety (grapefruit), 95
Osaka Purple variety (mustard green), 183
Osborne Prolific variety (fig shrub), 132
Outstanding variety (Romaine lettuce), 211
Oxheart variety (carrot), 170
oyster shell calcium, 24
Ozark Premier variety (plum), 124

Pablo Batavian variety (lettuce), 210
Painted Mountain variety (corn), 174
Pakistan variety (mulberry), 107
Palace de Versailles, Potager du Roi, 43*f*
Papri Sweet variety (pepper), 217
Parisian Pickling variety (cucumber), 191
Parmex variety (carrot), 171
Parris Island variety (Romaine lettuce), 211
parsley, 158
Parson Brown variety (orange), 101
parterre gardens, 43–45, 46–50, 59

pathways
   children's gardens, 55
   design considerations, 13
   parterre gardens, 44–45
   site evaluation, 6
   Starr residence garden, 59
   walled courtyard gardens, 66, 68
Pawnee variety (pecan), 89
peach trees, 116–18, 119*f*, 121–23, 247
pear trees, 12*f*, 18, 107–11, 249
peas
   children's gardens, 51
   cultivation guidelines, 18, 199, 201, 205–206
   pests summarized, 260
   varieties, 206–208, 240
pecan shell mulch, 33
pecan trees, 85–89, 246
Peking Black Crowder variety (southern pea), 208
Pencil Pod Black Wax variety (bush bean), 204
Pendolino variety (olive), 85
Pepperoncini variety (pepper), 216
peppers, 211–13, 215–17, 240, 262–63
perennial plants
   artichoke, 133–35
   asparagus, 135–37, 250
   bramble fruits, 137–41, 251
   grapevines, 141–46, 252
   Jerusalem artichoke, 146–47
   strawberries, 147–49, 252
pergolas, 18
perimeter fencing, 16–17
Perkins Mammoth Long Pod variety (okra), 226
permaculture gardening, 71–73
Persian Lime variety (lime), 98
Persian Mulberry variety (mulberry), 107
Persian variety (muskmelon), 193

persimmon trees, 77–80, 93–95, 111–14

Petroselinum crispum (parsley), 158

Phaseolus spp. (beans), 201–205

Philippine Lady Finger variety (okra), 226

Phil's Sweet variety (pomegranate), 116

pH levels
    site evaluation, 5–6
    soil preparation, 19–21, 24–25, 28

pH levels, requirements
    bramble fruits, 138, 140
    crucifers, 176
    legumes, 202, 205
    melons, 192
    nightshade family, 213
    See also specific plants, e.g.,
      blueberry shrubs; carrots; lettuce

phosphorus, 23–25, 95, 117–18, 173

Physalis ixocarpa (tomatillo), 211–13, 220–21, 262–63

Phytophthora. See diseases, overviews

pickling varieties (cucumber), 191, 240

Pierce's disease, grapevines, 142, 144–45, 146

pill bugs, compost heaps, 22–23

Pimpinella anisum (anise), 150

Pineapple variety (orange), 101

pine straw mulch, 32*f*, 33

Ping Tung variety (eggplant), 214

Pinkeye Purple Hull variety (southern pea), 208

pink plants, listed, 41

Pinus pinea (Italian stone pine), 77–81

Pisum sativum. See peas

pizza gardens, 64, 66, 67

planting guidelines, overviews
    melons, 187–89, 191–92
    trees, 77, 80, 94–95, 116–17
    See also specific plants, e.g.,
      blueberry shrubs; marjoram

planting guidelines, vegetables
    beet family, 162–63
    crucifers, 176
    cucurbits, 187–89, 195
    legumes, 199, 201
    nightshade family, 211–13
    onion family, 226–27
    See also specific vegetables, e.g.,
      corn

plant recommendations. *See* designing gardens, considerations; designing gardens, style approaches

plum trees, 116–18, 123–24, 247

pole beans, 18, 199, 201–204, 239, 260

Polish variety (amaranth), 165

pollination
    with flower plantings, 72, 152
    saving seeds, 236–38

pollination, special requirements
    shrubs, 127, 130–31
    trees, 86, 104, 107, 121, 123
    vegetables, 52, 170, 172–73, 189

pomegranate trees, 114–16

Ponca Butternut variety (winter squash), 197

Ponderosa variety (lemon), 97, 98

Ponkan variety (mandarin orange), 99

Poona Kheera variety (cucumber), 190

Potager du Roi, Palace de Versailles, 43*f*

potager gardens, designing, 43, 45–46

potassium amendments, 23–25, 117–18

potatoes, 211–13, 217–20, 240, 262–63

Powderblue variety (blueberry), 127

praying mantids, 243

precipitation, site evaluation, 10

Premier variety (blueberry), 127

Prescott Fond Blanc variety (muskmelon), 193

Principe Borghese variety (tomato), 223

Provider variety (bush bean), 204

pruning guidelines
    bramble fruits, 139, 140
    fruit shrubs, 126, 129
    grapevines, 143, 145–46
    trees, 77–80, 95, 118

Prunus spp., 116–24, 247

pumpkins, 198–99, 240, 258–59

Pumpkin Yam variety (sweet potato), 235

Punica granatum (pomegranate), 114–16

Purple Coban variety (tomatillo), 221

Purple Dragon variety (carrot), 171

Purple Italian Globe variety (artichoke), 135

Purple of Sicily variety (cauliflower), 182

Purple Passion variety (asparagus), 137

purple plants, listed, 41–42

Purple-Podded variety (pole bean), 203

Purple Sprouting variety (broccoli), 177

Purple Top White Globe variety (turnip), 187

Purple Vienna variety (kohlrabi), 185

Purple Viking variety (potato), 220

pyrethrum for insect pests, 242–43, 245

Pyro rootstocks (pear), 109–10

Pyrus communis (pear), 12*f*, 107–11, 249

Pyrus rootstocks (pear), 110

Queensland Blue variety (winter squash), 197

Quetzali variety (watermelon), 194

Quince rootstock (pear), 110

rabbiteye varieties, blueberry, 127

rabbit fencing, 17

raccoon fencing, 17

radishes, 175–76, 185–86, 241, 256–57

Rainbow Indian variety (corn), 174

Rainbow variety (Swiss chard), 169

rainfall, site evaluation, 10

raised beds, 25–26, 59

Ranger variety (peach), 122

Raphanus sativus (radishes), 175–76, 185–86, 241, 256–57

raspberry vines, 139–40

Rattlesnake variety (pole bean), 203

Red Acre variety (cabbage), 180

Red Burgundy variety (onion), 231

red cabbage varieties, 180

Red Cloud variety (potato), 220

Red Dale variety (potato), 220

Red Deer Tongue variety (lettuce), 210

Red Delicious variety (apple), 92

Red Finn Apple Fingerling variety (potato), 220

Red Giant variety (mustard green), 183, 184*f*

Red Globe variety (nectarine), 121

Red Globe variety (peach), 122

Red Gold variety (nectarine), 121

Red Gold variety (potato), 220

Redhaven variety (peach), 122

Red Kidney variety (bush bean), 204

Red mulberry species, 106

Red-Noodle Asparagus variety (pole bean), 203

red plants, listed, 42

Red Pontiac variety (potato), 220

Red Rubine variety (Brussels sprouts), 179

Red Sails variety (lettuce), 210

Redskin variety (peach), 122

Red Velvet variety (lettuce), 210

Red Wine Velvet variety (sweet potato), 235

Regal variety (sweet potato), 235

regional conditions, site analysis process, 7–11

residential gardens, design guidelines, 56–59

Rhizobium bacteria, 21, 31, 199, 206

Richmond Green Apple variety (cucumber), 190

Ringley's Porto Rico variety (sweet potato), 235

Rio Grande variety (peach), 122

Rio Red variety (grapefruit), 95

rock phosphate, 24

Romaine varieties (lettuce), 210–11

Romanesco variety (broccoli), 178

Romano variety (pole bean), 201*f*, 203

Roma varieties (tomato), 223, 241

rooftop gardens, 15, 61

rooms/houses, children's gardens, 51, 201

root rots. *See* diseases, overviews

rootstock recommendations

apple trees, 91–92

pear trees, 109–10

stone fruit trees, 116–17

Rosa Bianca variety (eggplant), 214

Rosborough variety (blackberry), 139

Rose Gold variety (potato), 220

rosemary, 158–59

Rosita variety (eggplant), 214

Rosmarinus officinalis (rosemary), 158–59

Rosseyanka variety (persimmon), 113

rotenone, insect pests, 243

Rotonda Bianca Sfumata di Rosa variety (eggplant), 214

rots. *See* diseases, overviews

Rouge d'Hiver variety (Romaine lettuce), 211

Rouge Vif d'Etampes variety (pumpkin), 199

Royal Chantenay variety (carrot), 171

Royal Oakleaf variety (lettuce), 210

Royalty Purple Pod variety (bush bean), 204

Royal variety (apricot), 120

Rubus spp., 18, 137–41

Ruby American Native variety (persimmon), 113

Ruby Perfection variety (cabbage), 180

Ruby Red variety (grapefruit), 94, 95–96

Ruby Red variety (Swiss chard), 169

Ruby Streak variety (mustard green), 183

Russian Red variety (kale), 183

Russian variety (pomegranate), 116

rust. *See* diseases, overviews

rye cover crops, 31

sage, 159–60

Sajo variety (persimmon), 113

Salad Bowl variety (lettuce), 210

Salvia officinalis (sage), 159–60

Sambucus spp. (elderberry), 128–30

Samdal variety (elderberry), 129

Sand Hill Preservation Center, 234

sandy soils, 19, 21

Sanguisorba minor (salad burnet), 152

sanitation practices

fig shrubs, 131

okra, 244

perennial plants, 139, 142, 144, 149

trees, 87, 116

See also diseases, overviews

San Pedro fig, 130

Santa Maria Pinquito variety (bush bean), 204–205

Santa Rosa variety (plum), 124

Satsuma variety (mandarin orange), 99

Saturn variety (peach), 123

saving seeds, 162, 236–38

savoy cabbage varieties, 180

Savoy Cross variety (kale), 183

scale considerations, designing gardens, 12–13

Scarlet Runner variety (pole bean), 203
Scarlet Sentinel variety (apple), 93
Schnittmangold Gelb variety (Swiss chard), 169
screening plants, listed, 40
Seascape variety (strawberry), 149
seaweed and molasses, 27, 28
seed saving, 162, 236–38
seed sources, listing of, 266–68
Seminole variety (pumpkin), 199
Sequoia variety (strawberry), 149
Serenade, fungal diseases, 242
Serrano varieties (pepper), 216
Seto variety (mandarin orange), 99
Seven Tops variety (turnip), 187
shade trees, 39, 80–89
shallots, 226–27, 231–32, 264
Shanghai variety (cabbage), 181
Shangri-La variety (mulberry), 107
Shanxi Li variety (jujube), 103
Sharp variety (sweet potato), 235
Shawnee variety (blackberry), 139
Sherwood variety (jujube), 103
Shinko variety (pear), 111
Shirey, Trisha, 46–47
Shogoin variety (turnip), 187
shrubs
    blueberry, 125–28
    elderberry, 128–30
    fig, 18, 52, 56, 130–32, 250
Sibley variety (winter squash), 197
Sierra Gold variety (muskmelon), 193
sight lines, site evaluation, 6
silver plants, listed, 42
Simpson Elite variety (lettuce), 210
single curtain system, pruning, 143
site evaluation
    location elements, 5–7
    map for, 7
    regional conditions, 7–11

size considerations, in designing gardens, 12–13
size element, in designing gardens, 38–39
slicing varieties (cucumber), 190–91, 240
slicing varieties (tomato), 241
Small Sugar variety (pumpkin), 199
SMR 58 variety (cucumber), 191
Smyrna fig, 130
Snowball Self-Blanching variety (cauliflower), 182
Snow pea varieties, 207
soaker hoses, 29–30
soil characteristics, site evaluation, 5, 9
soil preparation
    overview, 19–21
    amendments, 21–25
    raised beds, 25–26
    square foot gardening, 73
    See also specific plants, e.g.,
        blueberry shrubs; Italian stone pine
Solanum spp. *See* nightshade family
Southern Giant Curled variety (mustard green), 184
Southern peas, 207–208, 240
sow bugs, compost heaps, 22–23
soybeans, 199, 201, 205
Spaghetti Squash variety (winter squash), 197
Spanish Red variety (sweet potato), 235
spiders, 243
spinach, 162–63, 166–67, 241, 253
spinosad, insect pests, 245
Springer variety (spinach), 167
square foot gardening, 71, 73
squashes
    cultivation guidelines, 18, 187–89, 195

pests summarized, 258–59
summer varieties, 50*f,* 196–97, 241
winter varieties, 50*f,* 197–99, 240, 241
St. Anthony Marseilles variety (fig shrub), 132
Stanley variety (plum), 124
Star of David variety (okra), 226
Starr residence, 58–59
Steiner, Rudolph, 127–28
Stevenson's variety (sweet potato), 235
Stewart's Zeebest variety (okra), 226
stone fruit trees, 116–24, 247
storage requirements, site evaluation, 6–7
Straight Eight variety (cucumber), 190
strawberries, 147–49, 252
Strawberry variety (corn), 174
Strawberry variety (watermelon), 195
straw mulch, 32*f,* 33
structural elements. *See* architectural/ structural elements
Sugar Baby variety (watermelon), 195
Sugarlee variety (watermelon), 195
sugar snap pea varieties, 207
sulfur, 24, 126, 217–18, 246
summer squash varieties, 196–97, 241
Sumor variety (sweet potato), 235
sunlight, site evaluation, 5, 57
Supreme variety (Muscadine grape), 146
Surecrop variety (strawberry), 149
surround for insect pests, 246
Suyo Brocade variety (cucumber), 190–91
Sweet Banana variety (pepper), 217
sweet basil, 151
sweet bay, 151
Sweet Charlie variety (strawberry), 149
Sweet Chocolate variety (pepper), 217
sweet corn, 171–72

Sweet Meat variety (winter squash), 197–98

Sweet Passion variety (muskmelon), 193

sweet potatoes, 232–35, 265

Sweet Red Cherry variety (pepper), 217

sweet varieties (pepper), 216–17, 240

Sweet variety (pomegranate), 116

Swiss chard, 36f, 162–63, 168–69, 239, 253

Table Queen Bush Acorn variety (winter squash), 198

Tagetes lucida (Mexican mint marigold), 156–57

Tah Tsai variety (cabbage), 181

Taiwan Black-Seeded Long Bean variety (pole bean), 203

Tamopan variety (persimmon), 113

Tane-nashi variety (persimmon), 113

tangerine trees, 101

Tara variety (Muscadine grape), 146

Tatume variety (summer squash), 197

Tehama variety (mulberry), 107

temperatures
    compost heaps, 22
    mulching benefits, 31
    raised bed benefits, 25
    site evaluation, 8, 11

temperature tolerances. *See specific plants*, e.g., peas; persimmon trees

1015Y Texas Supersweet variety (onion), 231

Tendergreen variety (bush bean), 205

Tendergreen variety (mustard green), 184

tepees with plants, children's gardens, 51

Tete Noire variety (cabbage), 180

Texas A&M AgriLife... Laboratory, 25

Texas Blue Giant variety (fig shrub), 132

Texas Early Grano 502 variety (onion), 231

Texas Honeyjune variety (corn), 174

Texas Plant and Soil Lab, 25

Texas walnut trees, 89

Texroyale variety (peach), 123

Texstar variety (peach), 123

texture element, in designing gardens, 38

Thai basil, 151

Thai Green variety (cucumber), 191

Thai Green variety (eggplant), 214

Thai Hot variety (pepper), 216

Thai RW Tender variety (amaranth), 165

Thelma Sanders variety (winter squash), 198

Thessaloniki variety (tomato), 223

Thornless Mexican Lime variety, 98

Throgreen Lima variety (bush bean), 205

Thumbelina variety (carrot), 171

thyme, 45, 160

Tifblue variety (blueberry), 127

Tigertooth variety (jujube), 103

Tinda Gourd variety (summer squash), 197

Tisdale variety (apricot), 120

Tokinashi variety (radish), 186

Tokyo Bekana variety (mustard green), 184

Tokyo Market variety (turnip), 187

Tomate Verde variety (tomatillo), 221

tomatillos, 211–13, 220–21, 262–63

tomatoes, 18, 211–13, 221–23, 241, 262–63

Tomcot variety (apricot), 120

tool requirements, site evaluation, 6–7

topiary plants, listed, 40

Toscano variety (kale), 183

trees
    citrus fruit types, 12f, 18, 93–116
    city regulations, 60
    ornamental type, 18, 39, 89–93
    pest control summaries, 246–50
    planting/pruning guidelines, 77–80
    shade types, 39, 80–89
    Starr residence garden, 59
    stone fruit types, 116–24

trellises, 17, 141–42, 199, 201

Triamble variety (winter squash), 198

Tromboncino variety (summer squash), 197

Tropic Snow variety (peach), 123

Trucker's Favorite variety (corn), 174

True Platinum variety (corn), 174

t-tape irrigation, 30

turnips, 175–76, 186–87, 241, 256–57

20th Century Asian variety (pear), 111

UC varieties (asparagus), 137

umbrella plants, listed, 40

urban farms, 15

usage considerations, site evaluation, 6

USDA hardiness zones, site evaluation, 8

Utah Sweet variety (pomegranate), 116

utility lines, site evaluation, 6

Uzbekski variety (cucumber), 191

Vaccinium ashei (blueberry), 125–28

Valencia variety (orange), 101

Valencia Winter Melon variety (muskmelon), 194

Variegated variety (elderberry), 129

vegetable gardening
    overview, 161–62, 211–13
    pest control overviews, 250, 252–65
    saving seeds, 236–38
    See also designing gardens *entries*

vegetables, types
> beet family, 162–69, 239, 241
> carrots, 169–71, 239
> corn, 171–74, 239–40
> crucifers, 175–87, 239, 240, 241
> cucurbits, 187–99, 239, 240, 241
> legumes, 199–208, 239, 240
> lettuce, 208–11, 240
> nightshade family, 211–23, 240, 241
> okra, 223–26, 240
> onion family, 226–32, 240
> sweet potatoes, 232–35

Vernon variety (blueberry), 127
Verona variety (watermelon), 195
Vert Grimpant variety (muskmelon), 194
Victorian Red variety (Muscadine grape), 146
Vidrine's Midget Cowhorn variety (okra), 226
Vietnamese Red variety (amaranth), 165
views, site evaluation, 6
Vigna spp. *See* legumes
vines, listed, 39
Violet de Bordeaux variety (fig shrub), 132
Violetta Italia variety (cauliflower), 182
Violetta variety (sweet potato), 235
Viroflay Giant variety (spinach), 167
Vista White variety (loquat), 105
Vitis spp. (grape), 18, 45, 141–46, 252

Wakenda variety (sweet potato), 235
walkways, site evaluation, 6
walled courtyard gardens, 66, 68–70
walnut trees, Texas, 89
Waltham 29 variety (broccoli), 178

Wampum variety (corn), 174
Wando variety (English pea), 207
Warren variety (pear), 111
Washday variety (southern pea), 208
wasps, 243–44
water availability, in site evaluation, 5–6
water features
> children's gardens, 51–52, 54
> Japanese gardens, 62–63, 64
> in permaculture gardening, 71–72
watering guidelines, overviews
> compost heaps, 21
> trees, 94–95, 117
> vegetables, 188–89, 213
> *See also specific plants, e.g.,* blueberry shrubs; olive trees
watermelons, 191–92, 194–95, 241
weeds, 22, 23, 31, 33
weevils, pecan trees, 87–88
West India Burr Gherkin variety (cucumber), 191
Whippoorwill variety (southern pea), 208
White Bermuda variety (onion), 231
White Granex variety (onion), 231
White Half Runner variety (pole bean), 203
White Mountain Cabbage variety (collard greens), 182
White mulberry species, 106
white plants, listed, 42
White Satin variety (carrot), 171
White variety (pomegranate), 116
White Winter Pearmain variety (apple), 93
White Wonder variety (cucumber), 191
White Wonder variety (watermelon), 195

White Yam variety (sweet potato), 235
Wickson variety (plum), 124
wildlife, children's gardens, 52
Willamette variety (raspberry), 140
Willowleaf variety (sweet potato), 235
wilts. *See* diseases, overviews
wind, site evaluation, 5
Winter Bloomsdale variety (spinach), 167
Winter Density variety (lettuce), 208*f*
Winter Luxury Pie variety (pumpkin), 199
winter squash varieties, 197–98, 241
Womack, Larry Don, 88
Wonderful variety (pomegranate), 116
Wong Bok variety (cabbage), 181
Woodard variety (blueberry), 127

Yamato variety (cucumber), 191
Yard Long Armenian variety (cucumber), 191
Yard-Long Asparagus variety (pole bean), 201*f*, 204
Yellow Bush Scallop variety (summer squash), 197
Yellow Granex variety (onion), 231
yellow plants, listed, 42–43
Yellowstone variety (carrot), 171
Yok Kao variety (cucumber), 191
York variety (elderberry), 129
Yukon Gold variety (potato), 220

Zea mays (corn), 171–74, 239–40, 255
Zip Code Honey program, 61–62
Ziziphus jujuba (jujube tree), 101–103
Zuchinni Black variety (summer squash), 197